Health
and the
Female Adolescent

About the Editor

Sharon Golub is an associate professor of psychology at the College of New Rochelle and adjunct associate professor of psychiatry at New York Medical College. As both a nurse and a psychologist she has a long standing interest in the physical and psychological well being of the female adolescent. Dr. Golub is the editor of two other books, *Menarche* and *Lifting the Curse of Menstruation,* and has contributed many papers to scientific journals. She is currently editor of *Women & Health.*

Health
and the
Female Adolescent

Edited by
Sharon Golub

Health and the Female Adolescent was originally published in 1984 by The Haworth Press, Inc., under the title *Health Care of the Female Adolescent*. It has also been published as *Women & Health*, Volume 9, Numbers 2/3, Summer/Fall 1984.

Harrington Park Press
New York • London

ISBN 0-918393-05-1

Published by

Harrington Park Press, Inc.
28 East 22 Street
New York, New York 10010-6194

EUROSPAN/Harrington
3 Henrietta Street
London WC2E 8LU England

Harrington Park Press, Inc., is a subsidiary of The Haworth Press, Inc., 28 East 22 Street, New York, New York 10010-6194.

Health and the Female Adolescent was originally published in 1984 by The Haworth Press, Inc., under the title *Health Care of the Female Adolescent*. It has also been published as *Women & Health,* Volume 9, Numbers 2/3, Summer/Fall 1984.

Library of Congress Cataloging in Publication Data

Health care of the female adolescent.
    Health and the female adolescent.

    Reprint. Originally published: Health care of the female adolescent. New York: Haworth Press, 1984.
    "It has also been published as Women & health, volume 9, numbers 2/3, summer/fall 1984"
    Includes bibliographies.
    1. Adolescent medicine. 2. Adolescent girls—Health and hygiene. 3. Pediatric gynecology.
I. Golub, Sharon. II. Title.
[RJ550.H43 1985b]   616'.0088055   84-19772
ISBN 0-918393-05-1 (soft)

CONTENTS

Preface: Who Speaks for the Female Adolescent?   **xiii**
    Sharon Golub, PhD

The Challenge of Providing Health Care to Adolescents   **1**
    Susan M. Coupey, MD

Introduction   1
Common Health Problems of Adolescents   3
Providing Health Care to Adolescents   9

Nutritional Needs of the Female Adolescent   **15**
    Brian L.G. Morgan, PhD

Nutritional Needs of the Normal Female Adolescent   15
Needs of the Female Athlete   18
The Pregnant Adolescent   21
Common Nutritional Problems   23
Conclusion   27

**Reflections on Beauty as It Relates to Health
in Adolescent Females**   **29**
    Rita Jackaway Freedman, PhD

Introduction   29
Psychological Adjustment   30
Physical Health   35
Conclusion   42

**The First Pelvic Examination and Common Gynecological
Problems in Adolescent Girls**   **47**
    Karen Hein, MD

Introduction   47
Indications for the First Pelvic Exam   49
Proceeding Through the First Pelvic Examination   50

Sequential Events of Puberty 54
Common Gynecological Disorders 54
Conclusion 62

**The Effect of Pregnancy on Adolescent Growth
and Development** **65**
Phyllis C. Leppert, MD

Introduction 65
Physiological Changes During Puberty 67
Psychological Development 70
Physiological Changes During Pregnancy 71
Psychological Demands of Pregnancy 73
Comprehensive Prenatal Care 75
Summary 77

Scoliosis: Diagnosis and Current Treatment **81**
John B. Emans, MD

Introduction 81
The Natural History of Scoliosis 85
Documentation and Observation of Scoliosis 89
Non-Operative Treatment 90
Operative Treatment 96
Adolescent Kyphosis 98
Conclusion 99
Glossary 101

**Cigarette Smoking by Adolescent Females: Implications
for Health and Behavior** **103**
Ellen R. Gritz, PhD

Introduction 103
Effects of Cigarette Smoking on Health 104
Behavioral Aspects of Smoking in Women 105
Preventing the Onset of Smoking 109
Maintenance and Cessation of Smoking 110
Research and Policy Issues 112

The Troubled Teen: Suicide, Drug Use, and Running Away 117
Barbara Sommer, PhD

Suicide	117
Drugs	126
Running Away	133
Summary and Conclusions	137

Contributors

Susan M. Coupey, MD
Assistant Professor of Pediatrics
Albert Einstein College of Medicine
Director: Comprehensive Care Program for Chronically Ill
  Adolescents
Division of Adolescent Medicine
Montefiore Medical Center
111 East 210th Street
Bronx, NY 10457

Brian L. G. Morgan, PhD
Assistant Professor of Nutrition
Institute of Human Nutrition
Columbia University College of Physicians and Surgeons
701 West 168th Street
New York, NY 10032

Rita Jackaway Freedman, PhD
Psychologist, private practice
8 Overlook Road
Scarsdale, NY 10583

Karen Hein, MD
Director, Division of Adolescent Medicine
Assistant Professor of Pediatrics
Babies Hospital
Columbia University College of Physicians and Surgeons
630 West 168th Street
New York, NY 10032

Phyllis C. Leppert, MD
Assistant Professor of Obstetrics & Gynecology
Columbia University College of Physicians and Surgeons
630 West 168th Street
New York, NY 10032

John B. Emans, MD
Associate in Orthopedic Surgery
Children's Hospital Medical Center
Instructor in Orthopedic Surgery
Harvard Medical School
300 Longwood Avenue
Boston, MA 02115

Ellen R. Gritz, PhD
Director, Macomber-Murphy Cancer Prevention Program
Jonsson Comprehensive Cancer Center
University of California, Los Angeles
10920 Wilshire Blvd., Suite 1106
Los Angeles, CA 90024

Barbara Sommer, PhD
Lecturer
Department of Psychology
University of California, Davis
Davis, CA 95616

Preface:
Who Speaks for the Female Adolescent?

"Who speaks for the adolescent?" Dr. Victor C. Strasburger asked this question in a recent issue of the *Journal of the American Medical Association.* Contributors to this Special Issue of *Women & Health* seem to agree with his conclusion: We all do!

Contributors to *Health Care of the Female Adolescent* were asked to provide readers with some up-to-date information about the health problems adolescents commonly encounter and ways in which to prevent or treat them. Authors were also asked to share their insights and techniques in working with adolescents so that those of us involved in the delivery of health care services can learn how to be more effective in providing health care for this special population.

Susan M. Coupey reviews the major health problems seen among adolescents, emphasizing the differences between those affecting boys and girls. She notes, for example, that although accidents are a leading cause of death in this age group, they affect boys more often than girls. Teenage girls more often seek health care for problems related to sexual behavior and pregnancy. Coupey stresses the importance of psychological factors and risk taking behaviors and their relationship to the incidence of accidents and teen pregnancy. Prevention and health education are priority issues. And this theme is echoed in other papers as well. There is a need to teach adolescents practical skills that will enable them to resist pressure to engage in high risk behaviors such as the use of alcohol and drugs. Coupey also highlights the importance of health care programs specifically designed for teenagers that provide for their need for privacy in a setting that is comfortable for adolescents.

Adolescence is a time of rapid physical growth and there is an increased need for protein and iron, among other things, during these years. Brian L. G. Morgan reviews the nutritional needs of the female adolescent, including those of the female athlete. Pregnancy imposes additional, special nutritional demands. Morgan notes that the pregnant teen must have a diet that not only sustains her own growth but also provides for the growth of the fetus and later for lactation. And with dieting such an important part of the American culture, Morgan also discusses the increasingly common and very difficult to treat problems of obesity and anorexia in female adolescents.

Looks count. And the impact of the search for beauty can be seen on any newsstand. Rita Jackaway Freedman points out that adolescent girls are keenly aware of the importance of being pretty. Whereas boys learn to attract attention by acquiring skills, girls cultivate attractiveness in order to gain admiration and acceptance by others. Even with contemporary changes in the roles of women, the adolescent girl achieves status and identity by being attractive and finding a boyfriend. These pressures impose a great deal of stress on the adolescent girl and certainly influence her feelings about her body and her self esteem. Freedman describes the physical and psychological health hazards inherent in the adolescent girl's search for beauty, using as examples eating disorders, acne caused by makeup, and cosmetic breast surgery. The challenge to feminists who are concerned with women's health care is to help the female adolescent to adapt to societal norms while also doing what is best for her own physical and mental health.

Karen Hein reviews the indications for performing a pelvic examination in a teenager and the special approach that is best taken with the adolescent patient. She emphasizes the valuable patient teaching that can go on during the time spent with this patient, teaching that includes not only the presentation of factual information but also conveying an accepting and positive attitude toward one's body. Hein carefully details "how to do it" and this chapter can serve as an excellent guide for beginning practicioners who are working with female adolescents. She also reviews normal pubertal development and the common gynecological problems seen in this age group.

Among the ten million girls between the ages of 13 and 19, one million become pregnant each year. This accounts for 600,000 live births and 400,000 induced abortions. Phyllis C. Leppert addresses this health issue. She points out that in becoming pregnant the adolescent is hit simultaneously with three major adjustments. She must confront all of the developmental problems addressed by nonpregnant adolescents (increasing independence, separation from parents, new relationships with peers) and, in addition, she is forced to adjust to pregnancy and an intimate relationship with a member of the opposite sex. Leppert presents an overview of the problem of adolescent pregnancy, discussing its costs both to society and to the teenaged parent. She describes the physiological changes and psychological demands that pregnancy entails and addresses the adolescent's needs for comprehensive prenatal care.

Adolescent idiopathic scoliosis is an orthopedic problem affecting about 10,000 American adolescents, most of whom are female. It generally appears just before or during the teen years. The term scoliosis is derived from the Greek word meaning crookedness and in cases of scoliosis the spine curves from side to side, throwing the body out of alignment, causing physical deformity, and sometimes leading to impaired breathing or cardiac problems.

John B. Emans reviews normal anatomy of the spine as well as the deformities that may occur in cases of scoliosis. He describes techniques used in the diagnosis of scoliosis and then outlines the different treatments currently being used, including the newest nonoperative technique, electrical muscle stimulation, as well as bracing and surgery with spinal fusion. Emans notes that treatment of the adolescent with scoliosis requires a team approach in which the psychological and social needs of the adolescent who may face bracing or surgery are addressed. Professional psychological intervention or peer support groups may be helpful. Health professionals who are involved in counseling patients and their families may want to recommend "Scoliosis: A Handbook for Patients," a booklet available from the Scoliosis Research Society, 444 North Michigan Avenue, Chicago, IL 60611 (cost: $1.00). Also of great value is the book *Deenie,* by Judy Blume (1973, Bradbury Press). This novel, written for adolescents, describes the experiences, conflicts, and struggles of an early adolescent girl with scoliosis who must wear a brace. It provides the self conscious and frightened teen with some much needed explanation and reassurance in language that she can readily understand.

About three fourths of 12 to 17 year olds have smoked cigarettes. And, unbelievable as it may seem, cigarette smoking among teenage females now surpasses that seen among teenage males. Ellen R. Gritz describes the psychosocial factors that encourage young women to smoke and the disease consequences of tobacco use for women. Gritz believes that health care professionals and educators have an important responsibility in preventing smoking among these teenagers. She outlines psychological approaches that have proven to be effective in preventing the onset of smoking. These include focusing on the immediate, rather than the long-term physiological effects of smoking on the body and helping the adolescent to develop the behaviors she will need to resist the existing strong social pressures to smoke.

Suicide now ranks as the second most common cause of death among older adolescents. Barbara Sommer has focused on suicide and some of the other self destructive behaviors seen in female adolescents, specifically, drug abuse and running away. She compares the incidence of these behaviors in males and females. For example, more females than males attempt suicide but males are more likely to succeed. Sommer speculates that gender differences may play a role here since it is known that female adolescents are more willing to seek help. Sex differences in drug use are also noted: females are more likely to use stimulants than are males. This may be related to the use of amphetamines for diet control among the girls. Also, males drink more and, as noted above, females smoke more. Nonetheless, overall, Sommer points out, patterns of drug use for female and male adolescents are quite similar. She describes the predisposing factors associated with suicide, drug abuse, and running away and suggests that there are a number of ways to deal with the self destructive adolescent, ranging from prevention to treatment. All efforts aim at helping the young person to develop feelings of independence and competence.

Women & Health thanks the contributors to this volume for sharing their clinical and research acumen. Educators, nurses, physicians, psychologists, and others who work with adolescents will be able to draw upon the information presented here in order to better meet the health needs of the female adolescent.

Sharon Golub, PhD
Editor

The Challenge of Providing Health Care to Adolescents

Susan M. Coupey, MD

ABSTRACT. This article discusses the common health problems of adolescents with an emphasis on the differences between those affecting girls and boys. Statistics are given for mortality, hospitalization, and out-patient health visits from several different sources. Accidents and violence are the leading causes of death in adolescents but they affect boys much more often than girls. Teenaged girls are frequently hospitalized for health problems related to sexual behavior, mainly pregnancy and delivery related care and sexually transmitted diseases. Emotional and psychosomatic disorders are very common among adolescents of both sexes. Three general considerations important for the provision of health care to adolescents are discussed in detail. These include privacy and confidentiality, prevention of adverse health consequences of risk-taking behaviors, and the utilization of professionals from several different health care disciplines.

INTRODUCTION

Many professionals providing health care to adolescents find the experience difficult, time consuming and, at times, quite frustrating; in sum, a challenge. It is a challenge, however, that is not without its own excitement and rewards and those who meet it usually find the experience immensely satisfying. Adolescent health care is unique in that it demands of its practitioners the integration of the physiology, psychology, and sociology of each individual patient every time there is a health problem to be solved. With this age group, treating the "whole patient" is mandatory because of the very close interrelationships among physiology, behavior and social environment in the etiology and treatment of illness.

Many would argue that this is not a unique aspect of adolescent

The study upon which this paper is based was supported in part by a grant from the Robert Wood Johnson Foundation.

1

health care but applies equally well to the care of people in all age groups. Indeed, in the best of all worlds, with unlimited resources this would be the case. However, with adults, health care providers can usually achieve the necessary diagnostic and therapeutic result by directing their questions and treatment recommendations to the immediate problem, whereas with adolescents this is almost never sufficient to achieve the same results. For example, when a forty year old woman complains to her gynecologist of a vaginal discharge, that physician can perform the necessary pelvic examination, make the diagnosis and prescribe therapy in a thirty minute visit with the reasonable expectation that the woman will actually use the medicine and be cured. It would be nice and perhaps even helpful if the physician spent an additional thirty minutes discussing the woman's feelings about her sexuality but it is most often not necessary to do this in order to cure the vaginal discharge. With adolescents, on the other hand, it is almost always necessary to take a detailed psycho-social and sexual history, explore with the patient how she will pay for the medicine, role-play with her how she will tell her boyfriend and, in general spend as much time on the psychosocial aspects of the physical problem as is spent on the physical aspects. If this is not done and the teenager is treated in the same fashion as the forty year old woman, the likelihood of curing the vaginal discharge is much less.

In addition, there is such rapid developmental change occurring physically, psychologically, cognitively, and socially within every teenage patient that nothing can be taken for granted or expected to remain static over time. Because the patients themselves are continually developing and changing, the caregivers also must be adaptable and flexible. Adolescents can be conceptualized as sitting on a fence between childhood and adulthood and their health care providers must balance on that fence with them. They can be treated neither as children nor as adults and it is finding the appropriate combination of restrictive and permissive attitudes that is difficult for most caregivers.

Highly developed negotiating and mediating skills are a definite asset when dealing with this age group. Health care providers must find ways to encourage adolescents' independence without increasing their health problems or alienating their families. Simultaneously, providers must recognize and respond to teenagers' dependence without alienating or impeding their development. This is a delicate balance to achieve, and caregivers must constantly make fine adjustments in their attitudes towards and responses to both teenage patients and their parents.

Because physical maturity is completed several years before psychosocial and cognitive maturity, adolescence is a time of great vulnerability for accidents. Much of adolescent health care centers around treatment and prevention of health problems caused by risk-taking behaviors. The types of behaviors resulting in health problems are different for girls than for boys and thus the types of "accidents" suffered by girls are different but nevertheless highly significant.

In addition, adolescence is often a time of marked psychologic stress. Depression and other emotional and psychosomatic disorders are prevalent in both sexes in this age group. There is also a surprisingly large number of adolescents with various chronic illnesses, including asthma, sickle cell disease, sensory deficits, cancer and many others. Due to recent advances in medical technology and pharmacology, most of these chronically ill teens function reasonably well in their lives but they require continuous and regular contacts with the health care system.

A review of the common health problems of adolescents with an emphasis on the difference between those affecting girls and boys will serve as a preface to the discussion of some important considerations for the provision of health care in this population.

COMMON HEALTH PROBLEMS OF ADOLESCENTS

Any listing of the major health problems of adolescents in order of their importance depends upon the perspective by which such importance is being assigned. Different health issues emerge as leading causes of mortality as opposed to those which are precipitants of hospitalizations or reasons for seeking ambulatory care. In addition, there are regional, socioeconomic, and ethnic differences in the incidence of different diseases. From the perspective of mortality it should be noted that although the death rate for adolescents is relatively low as compared with other age groups, the teenage and young adult segments of the population represent the only groups for which that rate is currently accelerating. In 1976 there were approximately 15 thousand deaths in the United States among the 25 million young people between the ages of 12 and 17, representing a death rate of 6 per 10,000 (Kovar, 1979). The overwhelming majority of these deaths were from accidents and a significant minority from homicide and suicide. As can be seen in Figure 1, the kind of

FIGURE 1.   DEATHS FROM ACCIDENTS, POISONINGS AND
VIOLENCE IN THE AGE GROUP 1-24 YEARS
UNITED STATES 1976

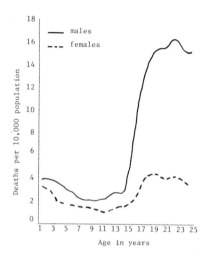

Adapted from Kovar, 1979

"accident" which is likely to cause death occurs much more frequently in adolescent boys compared with girls. Malignancy which
is more evenly distributed between the sexes, ranks as a distant
fourth behind these three traumatic causes of death. Indeed, 70% of
adolescents and young adults die from accidents and violence and
only 30% from diseases and conditions (see Table 1) (Kovar, 1979).

A review of the leading causes for hospitalization among adolescents reveals a similar prominence of conditions with an underlying
behavioral etiology. National data on hospitalization for 12-17 year
olds during 1976 reveal that 35% of hospital days for boys and 13%
for girls were for treatment of injuries and an additional 18% of hospital days for girls in this age group were for pregnancy and delivery
related care (Kovar, 1979). Much of this adolescent pregnancy requiring inpatient care can be conceptualized as resulting from an
"accident." Approximately 75% of teenage girls who carry a pregnancy of term report that the pregnancy was unintended and an even
higher percentage of teenagers who elect abortion have unintended
pregnancies (Alan Guttmacher Institute, monograph, 1981). In
Table 2, which lists diagnostic categories from our own adolescent

inpatient unit (Schonberg & Cohen, 1979), six percent of the admissions were for gynecologic disorders. These represent, in the main, girls with acute salpingitis another very common health problem resulting from sexual behavior. Very few adolescent boys require hospital admission for health problems related to sexual be-

TABLE 1   LEADING CAUSES OF DEATH IN 15-24 YEAR OLDS
IN THE U.S. 1978*

CAUSE OF DEATH	DEATHS PER 100,000 POPULATION
Accidents	60
Homicide	12
Suicide	12
Malignancy	7
All Causes	114

* Facts of life and death, U.S. Dept. HEW, National
Center for Health Statistics, DHEW Publication No.
(PHS) 79-1222.

TABLE 2   DIAGNOSTIC CATEGORIES OF 14,000 ADOLESCENTS
ADMITTED TO HOSPITAL 1967-1978*

DIAGNOSTIC CATEGORY	PERCENT OF TOTAL ADMISSIONS
Gastrointestinal Disorders	11
General Surgical Problems	11
Plastic Surgical Problems	7
Trauma	7
Neurologic Disorders	7
Cardiac Disorders	6
Pulmonary Diseases	6
Orthopedic Disorders	6
Endocrine-Metabolic Problems	6
Gynecologic Disorders	6
Tumors	6
Infections	5
Psychiatric Disorders	4
Hematologic Disorders	4
Drug Abuse	3
Other	5

Ref.: Schonberg & Cohen (1979) and unpublished data.

havior. On this particular adolescent unit, girls were not admitted for prepartum or postpartum care so the percentage of gynecologic admission is much lower than would be obtained for the total adolescent female population. If there is no change in current pregnancy rates for teens, four out of every ten girls who are now 14 will get pregnant in their teens (Alan Guttmacher Institute, monograph, 1981). Thus, fully 70% of adolescent deaths and a minimum of one-third of adolescent hospital days for both boys and girls are for health problems related to behavior. It is apparent that adolescent behaviors such as risk-taking, acting out, violence, and sexuality account for a significant portion of the morbidity and mortality in this age group.

Chronic illnesses, while accounting for few of the deaths in adolescents, with the exception of malignancies, are responsible for a large portion of both hospital admissions and outpatient visits among teenagers. An analysis of our own experience with 14,000 hospitalized adolescents, from 1967-1978 revealed that many of these patients were admitted because of chronic conditions (Schonberg & Cohen, 1979). While the sex ratio of total admissions was four to three with a male predominance, the specific diagnoses were not analyzed by sex (Table 2). Eleven percent of admissions were for gastrointestinal disorders (predominantly chronic inflammatory bowel disease, chronic liver disease, and eating disorders). Neurologic conditions (in the main seizure disorders), cardiac disease (both congenital and rheumatic), pulmonary dysfunction (primarily asthma), orthopedic deformity (scoliosis and slipped capital femoral epiphysis), endocrine disorders (diabetes mellitus and thyroid disease), and the plastic surgical repair of congenital defects were also frequent. Nearly half of all our adolescent hospital admissions were accounted for by these seven categories of chronic disease (Table 2). There are also large numbers of adolescents with chronic handicapping conditions such as hearing loss, mental retardation and cerebral palsy, who are not often hospitalized but who have special ongoing health care needs. It is estimated that about 5% of all 12-17 year olds are limited in activities because of their health.

The health problems of adolescents seeking outpatient care vary depending on the setting, the variety of services offered, and the population of teenagers served. Those professionals practicing in a rural setting will have a different experience and therefore a different perception of adolescent health needs than their colleagues in the inner city or at a tertiary care referral center. Beyond these perceived variations in adolescent health needs, many common health problems of teenagers never come to medical attention, in part, due to the obstacles to obtaining medical care that still exist for this age group.

Data from the 1975 Job Corps program for socially and economically disadvantaged youths aged 16-21 years (Hayman & Frank, 1979) revealed that about 2% of the 44,000 young people were terminated from the program for health reasons during the year. In spite of the fact that only 25% of Job Corps members were female, pregnancy accounted for 43% of the terminations. Mental and emotional diseases accounted for 32%, and accidents and violence for 4% of the terminations. The only other significant category of health problem associated with inability to continue the program was diseases of the nervous system and sense organs accounting for 5% of terminations (Table 3).

Our own experience with a general ambulatory care program for adolescents housed within a medical center complex in the Bronx, reveals a somewhat different but overlapping group of health problems (Schonberg & Cohen, 1979) (Table 3). Because the program includes gynecologic services, health problems in this area were frequently encountered. Venereal disease and contraception together accounted for about one quarter of all diagnoses among the 15,000 teenagers seen in a 12 year period. Five percent of the patients were without physiologic complaints and sought care for emotional disturbances, primarily, depression and anxiety.

TABLE 3   HEALTH PROBLEMS OF ADOLESCENTS:
                   A COMPARISON OF DATA FROM DIFFERENT SETTINGS

JOB CORPS TERMINATIONS[1]		HOSPITAL ADOLESCENT CLINIC[2]		HARLEM YOUTH SELF-REPORT[3]	
(N = 816)		(N = 15,000)		(N = 752)	
Pregnancy	43%	Venereal Disease	12%	Wearing Glasses	45%
Mental Disease	32%	Contraception	11%	Frequent Colds	25%
Nervous & Sensory Disorders	5%	Health Maintenance	9%	Skin Problems	22%
Accidents & Violence	4%	Infections	7%	Repeated Headaches	22%
Abnormal Signs & Symptoms	4%	Skin Problems	6%	Emotional Problems	21%

[1] Hayman & Frank, 1979

[2] Schonberg & Cohen, 1979

[3] Brunswick, Boyle, Tarica, 1979

Brunswick, Boyle, and Tarica (1979) interviewed black adolescents living in Harlem and compiled a list of self reported health problems in this population. These problems were quite different from those noted by health care providers and most of them went untreated. Wearing glasses was the most frequently reported problem with 45% of teenagers responding positively. Frequent colds or headaches, skin diseases, and emotional problems were reported by about one-fifth of these young people (Table 3). Some of these self-reported health problems were minor and did not require physician intervention. However, this study found that contrary to expectations, the seriousness of the health problem was not related to obtaining care from a physician. Girls reported more problems than boys but the utilization of physician care was not different between boys and girls. One of the most significant variables associated with a low level of physician care for an adolescent's self-reported problems in this racially and socioeconomically homogeneous population was having a mother who was born in the south. Those adolescents whose mothers were born in New York did not have fewer health problems but were more likely to get physician care for them. Since more than half of these families had medicaid and there was a large municipal hospital in the area, medical care was available to all regardless of ability to pay.

It is noteworthy that emotional and psychosomatic disorders emerge as major themes in all three of these diverse groups of adolescents in different health care settings. The incidence of psychosomatic complaints, depression, and self-destructive behavior is difficult to quantify and present in statistical form. However, every attempt to ascertain the frequency of such problems among young people seeking medical attention has reinforced awareness of the significance of emotional problems as a cause of adolescent ill health. More adolescent males die from suicide but non-fatal suicide attempts are far more common among teenaged girls.

Eating disorders are not mentioned in these surveys of adolescents from predominantly lower socioeconomic groups but they emerge as frequent problems in middle and upper middle class adolescents and are commonly encountered in suburban practices and college health services. The incidence of anorexia nervosa and bulimia is increasing and both these disorders have recently received much attention in the media. These psychosomatic syndromes affect girls almost exclusively and are responsible for considerable morbidity and health service utilization. Approximately three percent of girls with anorexia nervosa will die of their disease. Psychosomatic and emotional illness represent difficult therapeutic problems for the practitioner caring for teenagers and often require cooperation among several different health professionals for successful treatment. When acute illnesses that affect adolescents, such as influenza, infectious mononucleosis, acute appendicitis, gastroenteritis, and others are added to the above listing of behavioral, emotional, and chronic health problems, it becomes clear that teenagers have a significant degree of ill health and require outpatient, inpatient, and preventive care specifically designed to meet their needs.

PROVIDING HEALTH CARE TO ADOLESCENTS

Because of their special status in between childhood and adulthood and their unique vulnerability to illness caused by risk-taking behaviors, adolescents are better served by health care approaches that are tailored especially for them. In the design and execution of a health care program for teenagers whether it is a college health service, an acute care inpatient unit, a drop-in center, or a private office practice, there are certain key considerations that need be kept in mind. These include: the ability to provide private and confidential care, an emphasis on prevention of adverse health consequences of risk-taking behaviors, and the availability of appropriate multidisciplinary services in a setting which is comfortable for adolescents. Most adolescents are cared for by pediatricians and family physicians. There is, however, a subspecialty of pediatrics called adolescent medicine. Most of these specially trained physicians are occupied in the training of future pediatricians and in research on various aspects of adolescent health. Since there are only 40 programs each training one or two adolescent medicine specialists per year in the United States and approximately 40 million adolescents, most teenagers will continue to receive their health care from other primary care physicians.

The following comments are offered for consideration by all physicians who care for adolescents including pediatricians, family physicians, obstetrician-gynecologists, internists, orthopedists, and others. The first and perhaps most important need of the adolescent patient is the provision of private and confidential care. Clearly, not all health problems of teenagers require strict confidentiality, indeed most of the time these patients want their parents sympathy, support, and advice when facing the frightening prospect that their bodies may be malfunctioning. All teenagers do however, require privacy for at least part of the interview and physical examination. According time for a private interview and examination acknowledges the emerging independence and physical maturity of the adolescent. This private time allows for discussion and examination of embarrassing or secret behaviors, thoughts or parts of the body which are necessarily shared with a physician in the interest of good health care but are not necessarily shared in the same intimate detail with anyone else. At the conclusion of the interview and examination the positive findings should be summarized and shared with both the adolescent and when possible, the parents. A therapeutic plan which is acceptable to all can then be developed. Adolescents will often request that the physician keep confidential certain details of their behavior especially when this concerns sexuality and non-medical drug use. It is important that the health care provider respect this confidence in order to maintain trust and open communication with the teenage patient. However, dependent on the level of maturity of the adolescent and the seriousness of the health risk involved, the physician must, at times, insure that the parents are aware of the problems with their son or daughter and are motivated to seek appropriate help. It is best if this communication of sensitive health problems to the parents can be done with the adolescent's permission and in her presence or, even better, by the girl herself.

Let us consider for example, the case of a 14 year old girl who becomes pregnant. At this age in our culture, no adolescent either male or female is mature enough in either psychologic or socioeconomic areas to manage the emotionally-charged decisions or the complex and expensive logistics of preparing for the birth of an infant or an abortion without competent adult help. At the same time however, many 14 year old girls when faced with the shock of a diagnosis of pregnancy, will request that their family not be informed. Usually it is best to honor the request for confidentiality at that moment but to schedule a revisit within two days, explaining to the parents that you and the child are discussing a significant problem that needs to be investigated further. Often the child will tell the parents about the pregnancy during those two days of waiting and, if not, will be ready to have the physician help to tell them at the second visit. This way a breach of confidentiality on the part of the health care provider has been avoided and yet the teenager will have access to the adult support she desperately needs. This same health problem can affect a teenager at a different level of maturity however, and demands a different approach from the physician. For example, if pregnancy is diagnosed in a 17-1/2 year old college freshman who is living independently from her family, has a bank account, a part-time job and a supportive and mature boyfriend, it would seem inappropriate for the physician to insist that she share the diagnosis with her parents.

In the case of adolescent pregnancy as well as with many other health problems, the health care provider must maintain the trust and respect of the patient by giving private and confidential service while, at the same time, convincing the less mature teens with significant health problems to obtain the adult support necessary. For the more mature adolescents, physicians have a responsibility to explore the treatment plan in sufficient detail with the patient so that they are satisfied that it is sensible and safe and the resources are available to carry it out.

The second important consideration for adolescent health care is a strong emphasis on prevention of the major causes of morbidity and mortality in this population. Health education is an essential component of the care plan for all young people. This education, whether it is given in the schools, in programs such as the aforementioned Jobs Corps (Hayman & Frank, 1979) or by an individual physician to an individual teenage patient, should be geared to the developmental level of the adolescent and should focus on health risks that are meaningful to the teenager. Thirteen year olds are at very little risk of driving a motor vehicle while intoxicated with alcohol or marijuana, whereas this behavior is a major cause of health problems for many 16-19 year olds. The younger adolescent, on the other hand, will usually be at the stage of beginning experimentation with cigarettes or alcohol. They will respond to an education program which addresses immediate health risks of drinking or smoking at age 13 rather than health problems that only affect older teens or that will not appear for 30 or 40 years.

In addition, the type of education programs that attempt to teach practical skills for resisting pressure to engage in risk-taking behavior as well as accurate, up-to-date medical information seem to be more successful with teens. Some recent studies (see below correct reference form)[*] report a successful outcome with smoking and

---

[*]McAlister, Perry, & Killen, 1980; Perry, Killen, & Telch, 1980)

drug abuse prevention programs for high school students. Both these programs emphasized a positive, constructive approach to health as well as skills for fending off temptations to smoke or use drugs. The older prevention programs which relied on scare tactics and long lists of potential diseases caused by smoking, alcohol, or drugs were less successful. Individual health care providers should try to incorporate these techniques into their own counseling strategies for adolescents by trying to enhance their teenage patient's self-esteem and encourage positive health behaviors rather than chastising them for negative behaviors.

The physician-patient interview is also an excellent, safe place to practice skills for resisting pressure to engage in behaviors that are likely to lead to health problems. For example, a strategically posed question inquiring whether her boyfriend has his driver's license interjected in the middle of a discussion with a 17 year old girl about drinking at parties, can begin an exploration of alternatives to drinking and driving. This discussion should not focus on how bad it is for 17 year olds to drink at parties but, instead, on how important it is for this particular girl and her boyfriend to remain healthy and to avoid the major hazard to their health from drinking which is the risk of an accident while driving in a motor vehicle.

In addition to educational and preventive interventions aimed at health problems associated with substance use, adolescents require accurate information and counseling in the areas of sexuality, weight and nutrition, stress and emotional adjustment, and physical fitness. As noted above pregnancy, sexually transmitted diseases, eating disorders, emotional disorders and injuries are all important causes of morbidity in young people. Adolescent girls bear the brunt of health problems related to sexual behavior and particular attention to preventive counseling starting at age 12 or 13 is important. In addition, malnutrition in American adolescents is found almost exclusively in girls with anorexia nervosa. More than 60 percent of teenage girls have followed a weight loss diet at some time during their high school years. Thus, every adolescent girl should be interviewed routinely concerning her dietary habits.

The third important area of consideration for adolescent health care is the establishment of a setting in which teenagers feel comfortable and have access to different types of health professionals such as nutritionists, psychotherapists, educators, recreational therapists, and social workers as well as physicians. Within major medical centers and programs specifically designed for young people, all of these health professionals can be assembled in one place at one time. Within a private practice setting, access to health professionals from varied disciplines can be obtained through the use of consultation and community resources. Besides the obvious goals of adolescent health care to restore physical and psychological health and prevent future illness, there is the additional goal to shelter ongoing normal developmental processes and facilitate progression towards adulthood. For example, when teenagers must be hospitalized for several weeks because of an illness or injury, they should not be totally deprived of their education or peer group social interactions during this time. Adolescents have only a few years to learn all the skills necessary to become competent adults and retardation of essential developmental areas is a heavy price to pay in addition to illness.

The best way to protect adolescent development in spite of hospitalization, is to group adolescents together in one physical area and to assemble a multidisciplinary team of professionals skilled at working with this age group. This strategy, while certainly advantageous for the adolescent, is often more difficult and costly to administer. When ill teenagers are scattered in among geriatric patients or younger children, their specific and unique needs are not so obvious and therefore usually not addressed. This will save money in the short term but can be harmful especially to the chronically ill adolescent who may experience repetitive absence from school, become socially isolated and less productive as an adult.

One of the major challenges of adolescent health care is the necessity for professionals trained in quite different disciplines to come together and develop common lines of communication in the interest of understanding and treating the complex interactions of physiology, psychology, cognitive processes, and social environment that lead to ill health in teenagers. The rewards for engaging in this process are many: expanding professional horizons, influencing social change, and last but not least, helping to ensure that the next generation is able to develop into healthy, happy, and productive adults.

REFERENCES

Alan Guttmacher Institute, Monograph (1981). *Teenage pregnancy: The problem that hasn't gone away.* New York, NY: The Alan Guttmacher Institute.
Brunswick, A.F., Boyle, J.M., & Tarica, C. (1979). Who sees the doctor? A study of urban black adolescents. *Social Science and Medicine, 13A,* 45-56.

Hayman, C.R., & Frank, A. (1979). The jobs corps experience with health problems among disadvantaged youth. *Public Health Reports, 94*(5), 407-414.

Kovar, M.G. (1979). Some indicators of health-related behavior among adolescents in the United States. *Public Health Reports, 94*(2), 109-118.

McAlister, A., Perry, C., Killen, J., et al. (1980). Pilot study of smoking, alcohol and drug abuse prevention. *American Journal of Public Health, 70*, 719-721.

Perry, C., Killen, J., Telch, M., et al. (1980). Modifying smoking behavior of teenagers; A school based intervention. *American Journal of Public Health, 70*, 722-725.

Schonberg, S.K., & Cohen, M.I. (1979). Health needs of the adolescent. *Paediatrician, 8* (Suppl. 1), 131-140.

Nutritional Needs of the Female Adolescent

Brian L.G. Morgan, PhD

ABSTRACT. This paper emphasizes the need for adequate nutrition in order to sustain normal growth in adolescence. The special needs of the adolescent athlete and the pregnant teenager are also reviewed. Adolescent athletes require added nutrient intake for optimal performance. And, if pregnancy occurs, it imposes severe nutritional demands on the adolescent which must be met to prevent low birthweight infants. Although many teenagers consume a good deal of their food outside of the home, there seems to be little evidence of widespread nutritional deficiencies except with respect to iron. However, there is a high incidence of obesity and anorexia among female adolescents and these disorders are discussed.

NUTRITIONAL NEEDS OF THE NORMAL FEMALE ADOLESCENT

Adolescence is a time in the growth cycle when the body is growing faster than at any other period except during the fetal phase and the first year of life. Rapid growth begins around ten in the female and is usually completed by about fifteen, although there is considerable variation among girls as to the onset, rate and duration of the growth spurt. Obviously, someone on the twenty fifth percentile for height will grow less in absolute terms during her growth spurt than someone on the seventy fifth percentile. The increase in lean body mass associated with this growth, results in an increased need for protein, iron, zinc and calcium. Menstruation further increases the iron requirement. If other factors are superimposed on the female adolescent at this time, such as pregnancy, illness, or participation in competitive sports, her nutritional requirements will be further increased.

Nutrition and Normal Development

Growth rate in adolescence is related to the stage of sexual maturation and not to chronological age. This means that the most convenient way to assess the nutritional requirements of an adolescent is to judge where she is along the maturational curve at any given time using Tanner's (1962) classification of maturity. This is based on changes in secondary sex characteristics, namely breast changes and pubic hair growth. As shown in Table 1, a rating of 1 is prepubertal and of 5 is adult, although some additional growth usually occurs after this rating is reached.

Measurement of skeletal growth is a more direct and accurate way to assess physical growth but it necessitates radiographs. Since secondary sex characteristics are highly correlated with skeletal age, it is appropriate to use Tanner's system. For instance peak height velocity in girls occurs most often between pubic hair rating 2 and breast rating 3. Menarche usually occurs at rating 4 for pubic hair and breasts.

The National Academy of Science's Recommended Dietary Allowances for girls from 11-14 years are 2400 k calories[*] decreasing

Table 1

SEX MATURITY RATINGS[a]

Stage	Pubic Hair	Breasts
1	Preadolescent	Preadolescent
2	Sparse, lightly pigmented, straight, medial border of labia	Breast and papilla elevated as small mound; areolar diameter increased
3	Darker, beginning to curl, increased amount	Breast and areola enlarged, no contour
4	Coarse, curly, abundant, but amount less than in adult	Areola and papilla form secondary mound
5	Adult feminine triangle spread to medial surface of thighs	Mature, nipple projects, areola part of general breast contour

[a]Adapted from J. M. Tanner, Growth at Adolescence, 2nd ed., Blackwell, Oxford, 1962.

---

[*]Kilocalorie or Calorie (with a large C) is the term used to express the energy value of food.

to 2100 k calories for ages 15-18. Obviously these are average figures and any given individual may need more or less. Protein requirements, like the other nutrient needs, are greatest when growth is fastest and lowest when growth is slowest. The protein requirement for girls ranges from 50-55 grams according to this dictate. Calcium requirements are set at 1200 mg per day during the active growth period. Other essential minerals for growth are phosphorus, magnesium, copper, fluorine, sulfur, zinc, selenium, cobolt, molybdenum, nickel, tin, chromium, iodine and silicon. If the calcium and protein requirements are met by an ordinary varied diet, these other minerals should be present in adequate amounts.

Dietary Habits of Adolescents

By age 12 almost one third of all children in America eat only one meal each day with the family. By 17 almost half of the children eat just one meal at home. However, despite this, teenagers seem to cope quite well. Nutrition surveys (DHEW, 1970; Karp, Williams, & Grant, 1980; Schorr, Sanjur, & Erickson, 1972) show that girls on the whole have adequate intakes of calories, protein, vitamins, and minerals with the possible exceptions of iron and calcium. As dairy products have fallen out of favor among teenage girls it is not surprising that calcium intakes are below par. Girls from ages 9-11 consume an average of just two cups of milk per day and young women from 18-19 even less, 1-1/2 cups per day (Sanjur, 1979). Milk has been replaced in many teens' diets by soft drinks, which because of their high phosphorus content have further impaired the girls' calcium status. Vegetarian diets have become very fashionable among young adolescents and present a problem because they remove meat from the diet; meat is a source of easily absorbed iron and Vitamin B_{12} that must be obtained elsewhere when meat is not eaten.

Iron Deficiency

Iron nutriture requires special consideration during adolescence because iron is needed for growth. At puberty an adolescent female gains about 30 kg in weight as, over a seven year period, she doubles her weight. The growth spurt, coupled with menstrual losses, greatly increases the iron requirements of a young women to 1.6 to 1.85 mg of iron per day which represents 16 to 18.5 mg of dietary iron.

The larger the amount of iron ingested, the greater the amount absorbed. However, the efficiency with which iron is absorbed depends upon its source. For example, about 10-15% of the iron present in meat is absorbed, as compared to 5% from vegetable sources. Most adolescents consume from 10-15% of their diet as protein, mainly in the form of meat, which should meet the need for iron. The body's physiological state also has a major effect on iron absorption. Absorption is highest during periods of rapid growth when the body's iron stores are relatively low. Iron absorption is independent of age. However, it is linearly related to weight gain on a percentage basis.

In the event of pregnancy, another 400-1000 mg of iron is required to satisfy fetal demands, increased maternal red blood cell volume, as well as blood lost at delivery. The lactating adolescent loses 0.5-1.0 mg of iron per day. As iron stores are less than 350 mg in two thirds of young women, maternal anemia is a real danger unless supplemental iron is given at this time.

When an adolescent's diet is deficient in iron, she is vulnerable to iron deficiency anemia. Early signs of anemia include pallor, irritability, anorexia, listlessness, and fatigue.

NEEDS OF THE FEMALE ATHLETE

More than 6,000,000 high school students and several million younger adolescents in junior high school are regularly involved in competitive sports. This activity can have a very beneficial effect on development during adolescence both psychologically and physically. Often overlooked is the fact that dietary practices are an integral part of the success or failure in sports at any age.

Energy required for sports activities is supplied through aerobic and anaerobic metabolism of carbohydrate available as glycogen stores in muscles and the liver. This store is limited to 1000-1500 kcal and so must be replenished at regular intervals throughout the day. Involuntary weight loss indicates that more energy is being expended than is being consumed. This is the most common nutrition-related problem in young athletes (Smith, 1982). A regular mixed diet including adequate energy to meet the demands of training will provide all the necessary nutrients for optimal athletic performance. Vitamin, mineral, and/or protein supplements in no way enhance performance. The only exception to this is iron. Studies show that 10-20% of menstruating women in the United States are deficient in this nutrient (Cook, Finch, & Smith, 1976). Any form of anemia will of course compromise athletic performance. The level of body fat is a key factor in performance in competition sports (Smith, 1982). Excess body fat reduces speed, limits endurance, and does nothing to improve strength. Reducing body fat to the minimum compatible with fitness is the ideal for athletes competing in sports such as distance running, gymnastics, figure skating, and in sports categorized into weight classes such as wrestling, light weight rowing, and weight lifting. In this way the athlete has the maximum strength, endurance, and quickness for every unit of body weight.

To arrive at a pinnacle of fitness an athlete needs a well-planned program of energy expending exercise and a diet that will supply sufficient energy for training to prevent muscle being used as gluconeogenic fuel (Smith, 1976). If fat needs to be lost, the amount must be carefully calculated using skinfold calipers and comparing the given value to the ideal fat level for the chosen event. Two pounds per week is the ideal rate of fat loss in an adolescent athlete who is attempting to achieve peak fitness. This means a 1000 kcal per day negative energy loss. Energy expenditure is raised in training so that the desired energy loss occurs while the athlete maintains an energy intake of no less than 2000 kcalories per day. Once the target level of fitness is arrived at, the energy intake is increased enough to keep the competing weight at the optimal level but at the same time satisfy the energy demands of normal training and growth.

Different types of sports demand satisfaction of their energy needs in different ways (Finch, Miller, & Inamadar, 1976). Sporting activities involving short spurts of energy such as sprints, gymnastics, pole vault, and the long jump use energy derived from the muscle stores of ATP and phosphocreatine. During the competition adequate nutrient intake may be obtained from regular fluid intake and small high carbohydrate meals.

Sports like running games, rowing, and wrestling all require more intense energy output over long periods of time which is supplied both aerobically and anaerobically. For optimal performance a diet high in carbohydrate and low in both residue and salt intake should be taken 3-4 days prior to competition (Table 2). If the competition lasts over fairly long periods regular fluid intake and small high carbohydrate meals should be consumed during the competition (Smith, 1982).

A good state of hydration is essential for efficient energy metabolism. Water intake is the best means of satisfying the fluid needs of the athlete including losses through sweating. Sufficient water needs to be consumed to maintain the precompetition weight. Minerals and electrolytes lost in sweat are replaced with a mixed diet containing adequate energy to satisfy the needs of the active athlete. Concentrated beverages and salt tablets are unnecessary and merely make a given state of dehydration worse. This includes "athlete drinks" containing glucose and electrolytes, which have the added problem of eliminating the sensation of thirst and discouraging the much needed fluid intake (Smith, 1982).

The higher the level of glycogen in the muscle the better the performance (Hultman, 1978). Maximum glycogen levels can be achieved by 3 to 4 days of intense training of the muscles to be used in competition while eating a very restricted carbohydrate diet. This is followed up by 3-4 days of very reduced training activity while taking a diet providing 1000-1500 kcal. of carbohydrate. This 8 day plan maximizes the glycogen content of the muscles to be used. This should not be followed as a regular training regime and should only be used in preparation for two or three major competitions a season.

Table 2. Menu providing 3000 k calories and 125 g of protein, which
          is suitable for an adolescent athlete in rigorous training.

Breakfast (600 k calories)

1/4 cup low-fat milk
3/4 cup oatmeal
2 slices whole wheat bread
2 tsp margarine or butter
1 cup citrus sections
1 poached egg
8 oz apple juice

Lunch (870 k calories)

1 hamburger roll
4 oz lean beef patties
12 oz orange-apricot juice
1/2 cantaloupe melon
1 cup cream of tomato soup
8 crackers

Mid-Morning or Mid-Afternoon
(260 k calories)

6 dried apricot halves
6 cashews
8 oz orange juice

Dinner (930 k calories)

1 cup rice with herbs and mushrooms
1 tbsp butter or margarine
1 small tomato (sliced)
1 tbsp French dressing
3 oz roast turkey with green pepper
1 cup spinach
1 cup ice cream

Snack (340 k calories)

2 Dutch-twisted pretzels
12 oz tangerine juice

Of all girls trying out for competition sports, 8-10% have tissue iron depletion with low levels of plasma ferritin and transferrin saturation, although their hemoglobin levels will be in the normal range. Recent studies have shown that this compromises athletic performance, leading to low exercise blood lactate levels. Striking improvement in performance occurs in response to 10 days of iron therapy (Smith, 1982).

THE PREGNANT ADOLESCENT

Compared to births among the adult population there are a larger number of low birthweight infants (< 2500g) among the adolescent population and a smaller number of babies over 2500g. As mortality rates are much higher among low birthweight infants, this is not a very favorable situation. There is also a higher incidence of premature delivery, cesarian section due to cephalopelvic disproportion, toxemia, and anemia among adolescents. Many factors contribute to this unfavorable record, including low pre-pregnancy weight, low weight gain during pregnancy, smoking, the use of narcotics and the use of alcohol. In mothers under 15, an additional factor seems to be the gynecological age or years since menarche. The younger the gynecological age, the greater the risk of toxemia and the shorter the gestation period (Rosso & Lederman, 1982a).

Maternal nutritional status has an important effect on the outcome of pregnancy. Food intake tends to be extremely low (average 1700 kcal/day) compared to values for older women (Lunell, Persson & Starky, 1969). Several key nutrients are low in pregnant adolescents including calcium, iron, and vitamin A.

Weight gain during pregnancy has a positive influence on birth weight in pregnant women (Niswander, Singer, Westphal, & Weiss, 1969). Optimal fetal growth will only be achieved if the mother reaches approximately 10% excess body weight over ideal body weight at delivery (Rosso & Lederman, 1982b) which is about 25-30 lb. More weight gain has no effect and less has a negative influence. These guidelines apply to women over 20 and do not allow for growth of the adolescent.

There is little data on the nutritional needs of adolescents during pregnancy. Most of the recommendations come from extrapolating from the needs of adults with an allowance made for the change in growth requirements (Table 3). A good approximation is the dietary recommendation for the non-pregnant adolescent plus the increment

Table 3. Outlines the recommended requirements for selected nutrients during pregnancy in the adolescent and the most important sources of these nutrients.

TABLE 2

Nutrient	RDA	Best Sources
Calories	2400	meat, fish, poultry, fats, fruits, grains, legumes, nuts
Protein	78 g	meat, fish, poultry, eggs, milk, milk products, legumes/grains
Calcium	1200 mg	milk (all forms), yogurt, cheese, leafy green vegetables, clams, oysters, almonds
Iron	18 mg[a]	liver, meats, fish, poultry, whole grain and enriched cereals and breads, legumes cooked in an iron pan, leafy green vegetables, dried prunes, apricots and raisins
Folic Acid	800 mcg[a]	liver, yeast, leafy green vegetables, legumes, whole grains, fruits, and assorted vegetables
Pyridoxine (B$_6$)	2.5 mg[a]	wheat germ, meat, liver, whole grains, peanuts, soybeans, corn

[a]Supplementation with these nutrients is recommended as follows:

Iron-30-60 mg per day;
Folic Acid-400-800 mcg per day;
Pyridoxine-3-6 mg per day.

recommended for pregnant women. In many cases, especially among poor socioeconomic populations, nutrition prior to pregnancy is less than ideal. Such deficits must also be compensated for during pregnancy.

Calorie needs and weight gain in pregnant adolescents should be evaluated in terms of the individual's current body weight. Very lean girls need a higher than average weight gain. Obese girls also need the normal weight gain during pregnancy. This is not a desirable time for weight loss as fetal birthweight will be lowered by dieting (Huenemann, Shapiro, Hampton, & Mitchel, 1968).

Specific nutritional needs are summarized as follows:

Protein. An increase in protein intake of 30g per day is recommended for pregnancy (NRC, 1980). In addition a 2g allowance per day for growth is required. Protein intake among pregnant teenagers usually exceed these levels.

Iron. Many teenage girls suffer from iron deficiency. Hence, iron supplements are prescribed for all pregnant teens to ensure adequate iron nutriture (NRC, 1980).

Calcium. Adolescent populations have abnormally low calcium intakes (Public Health Report, 1969) and so supplements are often necessary to reach the 1600 mg target for growing pregnant teenagers.

Vitamin A. Vitamin A status has often been found to be below normal in teenagers and, being a key nutrient in growth processes, must be carefully maintained (Rosso & Lederman, 1982a).

Water Soluble Vitamins. Low intakes of riboflavin, pantothenic acid, thiamin, folic acid and vitamin C have been reported in teenagers, possibly due to a low intake of fresh fruit and vegetables. Adequate intake or supplements are needed to ensure optimal fetal growth (Rosso & Lederman, 1982a).

Other Nutritional Factors. Alcohol, smoking, and drug usage all have a tremendous impact on nutrition status and must always be considered when making up a diet for a pregnant teenager.

COMMON NUTRITIONAL PROBLEMS

Fad Diets

In many instances, unconventional or self-styled dietary regimens such as Zen macrobiotics or vegetarianism may be used by the adolescent as a means of creating some kind of self identity. The former can be disastrous as it will lead to undernutrition and growth retardation and the latter can lead to the same end result unless monitored carefully.

Adolescent Obesity

Obesity is not uncommon among adolescent women. The point at which an adolescent becomes sufficiently overweight to be called obese is difficult to define. Fat levels of the body change with age and with stage of puberty; therefore, the 40% or more fat criterion used for adults as the cut off value is inappropriate for adolescents. A figure of 20-25% above ideal body weight is widely accepted.

Fatfold thickness above normal is a good means of identifying an obese adolescent as opposed to one with an increased muscle mass or heavy body skeleton. For girls, this should be approximately 23mm at 14 years, 26 mm at 17 and 28 mm at 22 (Knittle, Timmers, & Katz, 1982). We have no way of identifying the obesity prone child. However, 12.5 to 30% of all teenage girls become obese, and 80% of these grow up to be obese adults. Obesity at any stage of life is associated with significantly increased health risks. In addition, long-term treatment of the condition has been disappointing and for the obesity-prone individual caloric restriction must be a life-long habit.

As energy and other nutritional needs parallel growth rate the adolescent has a higher need for any given nutrient in proportion to body size than people in other age groups except infancy. Caloric requirements in girls rise to a peak just after puberty at an average age of 12-13 and then drop in proportion to the reduction in growth rate (Heald, Daugela, & Brunschuyler, 1963; Wait & Roberts, 1932).

Several surveys have documented that obese adolescents eat fewer calories than many of their lean counterparts. They also tend to skip breakfast more often than normal weight teens, and overall eat fewer meats and snacks than lean teenagers (Hunemann, 1972). Thus not wanting to compromise their nutritional status further by an inappropriately designed diet, nutritional status of the patient should be determined before embarking on any dietary regimen. Overweight teenagers also tend to be less active than lean youths and therefore any weight loss program should include a planned exercise component.

Management of the obese adolescent must include a medical history and a thorough physical examination. Once all known medical causes have been ruled out, the patient is asked to keep a dietary history using behavior modification technique sheets as shown (Figure 1). This gives an evaluation of food intake, patterns of food intake, food preferences, degree of hunger, emotional problems, and the location of the major percentage of food consumed during the day.

The treatment of obesity is a complex issue and one can no longer merely think of the obese individual as one who over indulges because of a lack of self discipline. A simple caloric restriction may reduce the weight but cellular, metabolic, socioeconomic, cultural, and psychological factors all mitigate against maintenance of the reduced state. An obese teenager must be made to understand that a

Figure 1.

A typical behavior modification diary.

Food Eaten	Time	Social
Quantity Kind Calories No. of meals	When was the food eaten and how long did it take to eat?	In what context was the food eaten?

Location	Mood	Hunger
Where was food eaten?	State of mind when food was eaten.	0 - none 5 - ravenous

Adapted from: D.M. Shlian. Modifying behavior: From fat to lean. Patient Care, July 15, 1978, p. 239-244.

decrease in weight is not a universal cure all. Most of the psychological and social problems that she has attributed to obesity will remain, which might lead to some degree of depression and anxiety. The positive aspects of weight loss such as changes in body image and other medical and social benefits must also be emphasized.

Amphetamines are not of long term benefit in the treatment of obesity. The initial quick weight loss may interfere with the acceptance of effective weight maintenance therapies by misleading the person in regard to the nature and extent of effective long-term treatment. Using amphetamines to maintain weight loss is dangerous. Increasing doses have to be used which lead to addiction and extremely dangerous adverse effects on the brain. Long term use of amphetamines can produce an eight to ten point drop in I.Q.

Sufficient calories and protein must be supplied in the reduction diet of a teenager to allow for the growth and development of lean body mass while at the same time achieving a decrease in the degree of adiposity. Little is known of the interrelationship of cellular growth and development of the various organ systems. No one diet can be devised that will deny calories and protein to the fat depot while at the same time providing the necessary nutrients to the other tissues. Thus when restricting calories in the adolescent one must ensure that normal linear growth rates are maintained. Starting with the recommended daily allowance for calories and other nutrients, in which calories are distributed as 20% protein, 45% carbohydrate and 35% protein, calories are increased or decreased such that body fat (measured by skinfold calipers) is lost without compromising growth.

Anorexia Nervosa

The teenage chronic dieter is among the 60% of 12th grade girls who are or have been on a reducing diet (NCHS, 1976). This erratic starving and binge eating is clearly contraindicated in the average adolescent who usually has a marginal diet. Moreover, preoccupation with weight loss sometimes leads to anorexia nervosa which may be fatal.

Anorexia nervosa is a serious psychiatric disorder generally regarded as a disease of early to middle adolescence. The frequency of the disease has been on the increase during the past decade. Unfortunately, 1-2% die of the disorder (Hsu, Crisp, & Harding, 1979).

Anorexia nervosa patients are extremely varied. Some patients drastically reduce their caloric intake but continue to eat adequate quantities of protein. Others adopt very bizarre and idiosyncratic diets. Some even reduce water intake. Many abuse laxatives and diuretics and/or induce vomiting after eating. One extremely serious consequence of these practices is severe loss of body potassium. Potassium levels may drop to 50% below normal (Baker & Lyen, 1982).

The anorexic must be forced to eat and regain weight. Behavior modification and family support combined with psychotherapy can sometimes achieve remission of the anorexic symptoms in a few weeks (Minuchin, Rosman, & Baker, 1978). Patients with severe anorexia should be hospitalized until a stable weight consistent with life is attained. This may be 20-25% below the expected weight for age. In such patients the behavior modification and psychiatric support should be delayed until a stable weight is achieved.

The gastrointestinal system of these patients is normal (Silverman, 1977) and therefore the preferred mode of refeeding is by mouth. Medications and calories may be given by nasogastric tube or parenterally if food is totally refused. Special formulas are of no value. Tube feedings should provide maintenance calories (2000-2400 calories per day) and 50-55g protein per day. Most patients will do the best on a regular diet emphasizing easily absorbable carbohydrates. A caloric intake of about 67-75 kcal/kg/day and a protein intake of 50-55g per day is a reasonable level.

CONCLUSION

Because of their rapid growth and development, adolescents require special nutritional consideration. Athletic participation and pregnancy may add further physiological stress and females are particularly vulnerable to problems of obesity and anorexia. This paper has outlined the nutritional needs of the female adolescent and some ways in which they may be met.

REFERENCES

Baker, L., & Lyen, K.R. (1982). Anorexia nervosa. In: M. Winick (Ed.), *Adolescent Nutrition.* New York: Wiley.

Cook, J.D., Finch, C.A. & Smith, N.J. (1976). Evaluation of iron status of a population. *Blood, 48*(3), 449-455.

Department of Health, Education and Welfare. (1970). In: *Ten State Nutrition Survey 1968-1970.* Vol. V. DHEW publication no (HSM) 72-8133. Washington, D.C.: U.S. Government Printing Office.

Finch, C.A., Miller, L.R., & Inamadar, A.R. (1976). Iron deficiency in the rat. Physiological and biochemical studies of muscle dysfunction. *Journal of Clinical Investigation, 58*(2), 447-453.

Heald, F.P., Daugela, M., & Brunschuyler, P. (1963). Physiology of adolescence. *New England Journal of Medicine, 268,* 192-198.

Hsu, L.K.G., Crisp, A.H., & Harding, B. (1979). Outcome of anorexia nervosa. *Lancet, 1*(8107), 61-65.

Huenemann, R.L. (1972). Food habits of obese and nonobese adolescents. *Postgraduate Medicine, 51,* 99-105.

Huenemann, R.L., Shapiro, L.R., Hampton, M.C., & Mitchel, B.W. (1968). Food and eating practices of teenagers. *Journal of the American Dietetics Association, 53,* 17-24.

Hultman, E. (1978). Adolescent fitness. In: J. Parizkova and V.A. Rogozkin (Eds.), *Nutrition, Physical Fitness and Health.* Baltimore: University Press.

Karp, R.J., Williams, C., & Grant, J.A. (1980). Increased utilization of salty food with age among preteenage black girls. *Journal of the National Medical Association, 72*(3), 197-200.

Knittle, J.L., Timmers, K.I., & Katz, D.P. (1982). Adolescent obesity. In: M. Winick, (Ed.), *Adolescent Nutrition.* New York: Wiley.

Lunell, N.O., Persson, B., & Starky, G. (1969). Dietary habits during pregnancy—A pilot study. *Acta Obstetricia Gynecologicaet Scandinavia, 48,* 187-194.

Minuchin, S., Rosman, B., & Baker, L. (1978). *Psychosomatic families: anorexia nervosa in context.* Cambridge: Harvard University Press.

National Center for Health Statistics. (1976). *Monthly Vital Statistics Report—Health Examination Survey Data* Vol. 25#3, (NCHS publication No. HRA 76-1120).

National Research Council, Committee on Dietary Allowances. (1980). *Recommended Dietary Allowances,* 9th ed., Washington, D.C.: National Academy of Sciences.

Niswander, K.R., Singer, J., Westphal, M., & Weiss, W. (1969) Weight gain during pregnancy and prepregnancy weight. Association with birth weight of term gestation. *Obstetrics and Gynecology, 33,* 482-491.

Public Health Report. (1969). *Dietary habits of pregnant teenagers and their potential rela-*

tion to pregnancy outcomes. (U.S. Public Health Report No. 84). Washington, D.C.: U.S. Government Printing Office.

Rosso, P. & Lederman, S.A. (1982a). Nutrition in the pregnant adolescent. In: M. Winick (Ed.), *Adolescent Nutrition.* New York: Wiley.

Rosso, P. & Lederman, S.A. (1982b). Nutrition in the pregnant adolescent. *Current Concepts in Nutrition, 11,* 47-62.

Sanjur, D. Teenagers and nutrition. (1979). *The Professional Nutritionist, 11*(1), 1-5.

Schorr, B.C., Sanjur, D., & Erickson, E.C. (1972). Teenage food habits. *Journal of the American Dietetics Association, 61,* 415-420.

Shlian, D.M. (1978, July 15). Modifying behavior: From fat to lean. *Patient Care,* pp. 239-244.

Silverman, J.A. (1977). Anorexia nervosa: clinical and metabolic observations in a successful treatment plan. In: R.A. Vigersky (Ed.), *Anorexia Nervosa.* New York: Raven Press.

Smith, N.J. (1976). In: *Food for Sport.* Palo Alto: Bull Publishing.

Smith, N.J. (1982). Nutrition and the adolescent athlete. In: M. Winick (Ed.), *Adolescent Nutrition,* New York: Wiley.

Tanner, J.M. (1962). In: *Growth at adolescence,* 2nd ed., Oxford: Blackwell.

Wait, B., & Roberts, L.J. (1932). Studies in the food requirement of adolescent girls: I. The energy of well nourished girls 10-16 years of age. *Journal of the American Dietetics Association, 8,* 209-237.

Reflections on Beauty as It Relates to Health in Adolescent Females

Rita Jackaway Freedman, PhD

ABSTRACT. Because an attractive appearance is so essential to the feminine gender role, the search for beauty causes special adjustment problems for adolescent girls. Psychologically they suffer from negative body image, lowered self-esteem, and achievement conflicts. Physically their health is undermined by current beauty norms which foster eating disorders, cosmetic acne, and breast surgery. The effects of physical fitness programs, the role of the media, and the influence of changing gender roles are also discussed.

INTRODUCTION

There is a growing awareness among psychologists that physical attractiveness is a highly significant psychosocial variable, and that attractiveness stereotypes are stronger for females than for males (Wallston & O'Leary, 1981). Long before puberty, girls learn that beauty is a basic dimension of the feminine gender role. Cultivation of attractiveness consequently becomes a major task for adolescent females, one which affects both their physical and psychological health. The high cost of such historical practices as foot binding and tight corset lacing is obvious, whereas the more subtle effects of current beauty demands often go unrecognized. This paper examines some of the contemporary consequences of the search for beauty on the psychological and physical health of adolescent girls, including body image, achievement and affiliation conflicts, eating disorders, physical fitness, cosmetic caused acne, and breast surgery.

Adolescence is not a sudden distinct occurrence, but rather a continuation of earlier developmental processes. According to the cognitive model of gender acquisition, sex role identification begins before the age of two. Girls first form a rudimentary notion that they are a girl, begin to realize that girls look and act a particular way, and eventually conclude that they also should look and act like other girls (see Hyde & Rosenberg, 1980). Because femininity is so closely linked with beauty, this process translates into the equation: "I am a girl, girls are pretty, therefore I should also be pretty."

Being female is associated with beauty from the earliest days of a child's life. In a study where newborn boys and girls were carefully matched for equivalent length, weight, and Apgar scores, parents more often described their infant daughters as beautiful, soft, pretty, cute and delicate while they rated their sons as strong, larger-featured, better coordinated and hardier (Rubin, Provenzano, & Luria, 1974). Though preferences for having sons still persists, one of the most common reasons given by women who did prefer to have a daughter, was that it would be fun to dress her and fuss with her hair (Coombs, Coombs, & McClelland, 1975). When asked what kind of person they wanted their child to become, parents mentioned "being attractive" and "having a good marriage" far more often for daughters than for sons (Hoffman, 1975).

Through the long process of socialization, children learn that for males, the body is to be developed, strengthened, made more functional and competent; for females the body is to be preserved, protected, made more beautiful (Rindskopf & Gratch, 1982). A girl grows up playing with toy cosmetic kits, surrounded by constant models of beautiful women on billboards and on TV. She has heard the story of the ugly duckling who turns into a swan and "decides that beauty, like menstruation is something that happens to girls at adolescence. She will wait" (Stannard, 1971, p. 119).

The consequences of socializing girls to invest so heavily in their appearance will be considered first with respect to their psychological adjustment during adolescence, and then in relationship to their physical health.

PSYCHOLOGICAL ADJUSTMENT

Narcissism

Narcissism has been called the hallmark of adolescence. As new levels of cognitive reasoning are attained, adolescents paradoxically misinterpret social and body signals, leading to heightened egocentrism and narcissism. Adolescent thought is distorted by concepts such as the "imaginary audience" (that everyone else is constantly scrutinizing their behavior and appearance) and the "personal fable" (that their experiences are unique and immune from the consequences of ordinary reality) (Elkind, 1978). These two common characteristics of adolescent thinking can make young women more vulnerable to cultural demands for exaggerated beauty images. For example, the anorectic may use the imaginary audience to display her distorted body, while she employs the personal fable to deny her hunger pangs.

The narcissistic self-obsession of teenagers is part of their attempt to strengthen the developing ego. Freud (1965) also regarded narcissism as a "peculiarity of mature femininity." In his view, female vanity derived from penis envy, a compensation for women's "original sexual inferiority." He interpreted the "turning inward" of girls at puberty as evidence of weakness in their capacity to establish meaningful relationships. Gilligan (1982), on the other hand, argues that heightened narcissism among adolescent girls need not represent relational failure, but instead indicates a new responsiveness to the self, and thus is a healthy part of their search for a unique personal identity. Adolescents of both sexes may be equally narcissistic, but express it in somewhat different ways (Maccoby & Jacklin, 1974). Boys "show off" through status and power, thus signalling their potency; girls cultivate attractiveness, thus seeking admiration and indicating to boys their social interest.

Achievement and Affiliation

According to Douvan and Adelson (1966) there are two distinctive adolescent adjustment crises—the masculine and the feminine. Boys' conflicts center on achievement, autonomy and control, whereas girls' concerns focus on affiliation and interpersonal relationships. Girls say that acceptance by others and adult recognition are the qualities which make them feel important and useful. They are less likely than boys to name personal achievement as a source of self-esteem and compared to boys, their goals for the future are more ambiguous. Erikson (1968) concludes that "much of a young woman's identity is already defined in her kind of attractiveness": and he suggests that a girl must spend time in "an identity moratorium," waiting until "her attractiveness and experience" help her find a mate who will then shape her destiny. In his monumental study of youth in the 1950s, Coleman reported that the most salient lesson taught in the "underground high school curriculum" was the importance of physical attractiveness for girls (see Lott, 1981). Times have changed since these ideas were formulated, nevertheless, the double bind between achievement and affiliation remains a salient issue for adolescent girls. Lott asserts that the fundamental task facing today's adolescent girl remains unchanged; "to enhance her attractiveness and find a boyfriend . . . therefore she learns to smile a lot, be pleasant, nonassertive, well groomed, friendly, and available" (Lott, 1981, p. 82). She calls it the "put on a happy face phenomenon" whereby girls create a "cosmetic exterior of the self" in order to attract males and thereby achieve status and identity. Lott adds that these generalities are as valid to contemporary girls as they were 20 years ago, and apply to all social classes and ethnic groups.

Horner's work on achievement motivation suggested that young women are in conflict, fearing that they will be "unsexed by success" (see Hyde & Rosenberg, 1980). For example, after winning a so-called intelligence test against an opposite sex classmate, the scores of seventh grade girls dropped on the second half of the test whereas the boys' scores went up. The girls explained that they didn't like to beat boys in a competitive game, and "would rather be popular than have good grades or win against a boy" (Condry & Dyer, 1977). They also know that good looks are essential to attaining popularity. Studies confirm that attractive girls date more often than unattractive ones, whereas appearance is unrelated to dating frequency in boys (Berscheid, Dion, Walster, & Walster, 1971). High school boys placed good looks and good body as first and second factors in date selection, while girls listed intelligence as the most important characteristics sought in a boy friend.

Adolescent girls need a good deal of support in handling the achievement-affiliation conflict. When parents, especially mothers, tacitly grant permission *not* to be popular, it can protect a girl from uncertainty over body image, and free her to explore and develop her own special talents (Rivers, Barnett, & Baruch, 1979). Moreover, there is evidence that attitudes are changing as a result of changing sex roles. An exploration of the relationship between attractiveness, career orientation and course selection found that career orientation was positively associated with physical attractiveness. Contrary to the traditional stereotype of the ugly, brainy, bookworm, those girls who took traditionally masculine courses like physics and math were not perceived as unattractive (Lanier & Byrne, 1981).

Media and Body Images

Restructuring ones body image to conform to pubertal changes, and developing a healthy sense of body satisfaction, are major challenges of adolescent adjustment. In today's technological society, body image is highly influenced by mass media portrayal of ideal body types. Teenage girls are presented by the media in two extremes: either as extroverted free spirits packaged in artifice, or as introverted dolls cloaked in innocence. Umiker-Sebok (1981) reports that girls seem to undergo a second infancy in these ads, displaying soft hairless skin, pink cheeks, a wide-eyed unfocused gaze. They are often shown stroking stuffed animals or sucking on something. These childish qualities, which are also part of the feminine beauty mystique, serve to convey helplessness and are part of the allure needed to attract a boy for protection.

Cosmetic advertisements have been shown to affect the "conception of social reality" of high school girls (Tan, 1979). One single 15 minute exposure to a series of beauty commercials increased the degree to which 16-18 year old girls perceived beauty as being "important to their own personality" and "important to being popular with boys." The main thing these ads encourage a girl to do to contribute actively to her own development is to transform herself into a beauty object (Umiker-Sebok, 1981). Through such ads, achievement and beauty cultivation become confounded.

The adolescent girl is also a consumer. Giant business and media networks join in trying to define, create and cater to her "needs" for beauty products. Acquiring these beauty aids is a costly process that diverts a girl's meager resources into purchase of the latest fashion necessity. Analysis of such magazines as *Seventeen* and *Mademoiselle* indicates a preponderance of ads for personal care products which promise to transform an ordinary girl into something special (Umiker-Sebok, 1981). She is encouraged to put her best face forward but at the same time to hide behind an artificial mask. Ads which are designed to convince a girl that she must make-up and make-over to look acceptable eventually undermine her self-confidence and contribute to negative body image.

Accepting one's rapidly changing body is not an easy task. Early adolescence marks the highest degree of anxiety and the greatest dissatisfaction over body image (Petersen, 1979). College students of both sexes retrospectively rated their attractiveness during adolescence as lower than that during childhood or at the present time (Unger, 1981). While both sexes worry, their concerns take different forms. Boys focus on size and strength, variables that are related to power; girls' concerns are more numerous and varied, but focus on appearance (Frazier & Lisonbee, 1960). Because they have been socialized to overemphasize appearance as essential to femininity, girls have a more difficult time developing a positive body image.

A study of 20,000 adolescents by Offer, Ostrov, and Howard (1981) found that girls had more negative feelings about their bodies than boys. Over 60% of high school girls say they want to change their looks as compared with only 27% of boys (Douvan & Adelson, 1966). Boys tend to rate their own personal appearance as more desirable than that of their peers, whereas girls rate themselves as less attractive than other girls (Musa & Roach, 1973). While measures of personal adjustment in boys seem to be unrelated to their self-evaluation of appearance, girls in the low range of personal adjustment also tend to rate themselves as low in personal appearance. Hence self-esteem is associated with appearance in girls but not in boys.

The high degree of negative body image experienced by many adolescent girls is interpreted as a direct result of cultural preoccupation with female appearance (Clifford, 1971; Offer et al., 1981). Because they have a stronger need to feel attractive, and because they are often encouraged to seek personal identity through male attention, girls suffer greater insecurity over their changing bodies during puberty. Since good looks are also stereotypically associated with personal worth, an unattractive changing body translates into a lack of self-esteem. The connection between good looks and worthiness as a female may be so deeply ingrained that it can remain throughout a woman's life, making her insecure and dependent on others for approval of appearance.

Turning now to problems of physical health, the recent increase in eating disorders among adolescent girls clearly reflects a complex interrelationship between social, psychological and physical variables, and shows how a high need for approval coupled with insecurity over appearance can contribute to adolescent pathology.

PHYSICAL HEALTH

Eating Disorders

For the past 20 years, feminine beauty has been equated with ultra slimness. The dramatic thinning of the American woman has contributed to weight obsession and serious eating disorders, especially in adolescent girls. Among the secondary sexual characteristics that become dimorphic at puberty is a substantial difference in subcutaneous tissue. While the ratio of fat increases almost 100% in girls until it makes up 25-30% of body weight, the ratio in boys actually drops to 15% (Faust, 1979). A study of 10th graders found that 55% of girls compared to 13% of boys were concerned with weight, especially fat on hips, stomach and legs (Frazier & Lisonbee, 1960). Females in general tend to value smallness and also tend to overestimate the size of their bodies (Halmi, Goldberg, & Cunningham, 1977). Girls describe the ideal figure as being "smaller than you are in all dimensions except bust." The ideal female physique that is currently considered attractive is further from the physiological norm than the ideal male physique (Unger, 1979). Overweight girls as compared to boys are frequently judged differently by others. For example, a study of admission practices at prestigious colleges showed that the rejection rate for overweight girls was three times higher than for overweight boys with similar academic records (Canning & Mayer, 1966). Thus part of the sex difference in concern over fat may be due to cultural bias against fat females combined with actual physical differences in the body fat component.

By the end of high school, 60% of the girls have already been on a "serious" diet and about 30% are dieting at any one time. In contrast only 24% of boys had been on a diet and 6% were currently dieting (Dwyer, Feldman, & Mayer, 1967). Girls attribute their fat to different causes than boys and therefore seek different solutions. Overweight girls attribute fat to overeating and typically handle overweight by dieting. Boys feel they are heavy because they have excess bone and muscle rather than fat, and therefore regard excess weight as desirable or attempt to increase their exercise. Chronic dieting during adolescence is thought to slow down the metabolic rate, resulting in weight gain under normal caloric intake and contributing to life long weight control problems.

Anorexia nervosa exemplifies how cultural factors can cause pathology to take a particular form. Bruch (1973) describes anorexia as a caricature of what will happen when the belief that dieting will make you beautiful and happy is taken too literally. Anorexia is of course a highly complex disorder. The following is not intended to fully explain its etiology, but rather to indicate how an exaggerated emphasis on physical appearance in the lives of growing girls contributes to this syndrome.

As is well known, symptoms of anorexia begin almost exclusively among adolescent girls and young women. Described as a spreading epidemic, it now affects about one percent of the females between the ages of 12 and 25. The use of vomiting and purging to control weight has likewise increased. Because it often goes undetected for years, its frequency is uncertain, however it is estimated that about 20% of female college students have had some involvement with bulimia (Halmi, Falk, & Schwartz, 1981; Mayer, 1982). We have seen that girls typically experience a large increase in fat-ratio during puberty, for estrogens bind fat, particularly on breasts, thighs and hips, thus filling out the female form. While the pubertal changes of breast development and menstruation seem inevitable, body weight, in contrast, does appear to be potentially under voluntary control; hence girls turn to weight manipulation to reestablish a sense of mastery over their bodies.

During childhood girls have learned that looking pretty means being dainty, delicate, petite. In a sense, puberty transforms a girl into a woman without her consent; it betrays her by making her both more and less feminine at the same time. The hormones that inflate her breasts, also layer her thighs with "unsightly" fat, and cover her legs with "superfluous" hair. The size, contours, smells and texture of an adult woman contradict the soft, sweet childish aspects of feminine beauty standards emphasized in the media. Thus the anorectic feeels driven to simultaneously maintain the presence and absence of an adult female body.

According to the traditional psychoanalytic view, anorexia occurs when Oedipal urges resurface at adolescence, preventing healthy acceptance of the adult female role. Self-starvation is interpreted as a defense against incestuous urges and against fear of oral impregnation. In contrast, some contemporary theorists are suggesting that anorexia is not a rejection of womanhood at all, but just the opposite; a dramatic attempt to achieve ultra femininity. Bruch (1973) concludes that anorexia reflects an acceptance of the feminine ideal along with an exaggerated striving to achieve it. Girls are especially vulnerable because they have been taught to solve psychological problems through manipulation of the body as a characteristic feminine response. They strive to control their appearance in order to please others and thereby validate their own sense of worth. Levenkron (1982) asserts that anorexia does not start as a rejection of sexuality, but as a desire for the highest level of achievement. Anorectics are frequently hard working perfectionists. The dedication and self-discipline that brought rewards in childhood are used to create the perfect feminine body, and with it a sense of personal identity. Paradoxically, as they avidly pursue a feminine beauty ideal, the result is a desexualized, often infertile person.

While a young Victorian girl laced herself in whalebone to fashion the perfect figure, today's anorectic redesigns herself in a corset of self-control. Through super-human effort she becomes as thin as the skeletal models who display the latest Paris fashions. One model reported that it took a whole crew of people to get her into a pair of size 3 designer jeans and carry her into position to be photographed because she could neither bend nor walk. One reason that anorexia is almost non-existent among males, says Levenkron (1982), is because they are not exposed to emaciated masculine models nor subjected to extreme cultural demands for beauty and thinness.

The increase in eating disorders has also been linked to recent changes in gender roles. A girl now grows up believing she should take control of her life and '' become something,'' yet at adolescence she still gets conflicting messages about appropriate feminine behavior and appearance. As one former anorectic said, ''I feel my illness was a product of the conflicts all women have to deal with being female in our society today. There is such gross contradiction in the emphasis on being a powerful, full bodied, strong willed female, and on being skinny, fragile, and young looking'' (Combs, 1982).

Girls with eating disorders are trying to attract attention and to win acceptance through body transformation, just as most healthy girls also alter themselves through beauty rituals in their search for popularity. Treatment of eating disorders must include recognition of contributing social factors, especially the popular equation of beauty with thinness. Girls need greater awareness that the emaciated models contrived by the media are distorted exaggerations rather than healthy norms. To help them, the National Association of Anorexia Nervosa and Associated Disorders (Box 271, Highland Park, IL 60035) has established chapters on college campuses across the country. A theater group has been touring schools with a show called ''Food and Fright'' which challenges the formula that thinness = happiness, and which sensitizes girls to the realization that they are not alone in their eating disorders.

Physical Fitness

A century ago Susan B. Anthony wrote that women need strong bodies as well as quick minds in their struggle for equality. Physical fitness is related to beauty stereotypes and the new emphasis on athleticism is having both a positive and negative effect on the health of adolescent girls. By the end of puberty there is a 10% sex difference in muscle ratio, with muscles comprising about 36% of female body weight as compared to nearly 45% of male body weight. Females are typically 50% weaker than males in upper body strength and 10% weaker in lower body strength relative to body weight (Faust, 1979). Sex differences in exercise patterns also influence muscle development, and as girls engage in more strenuous and continuous athletic programs, the magnitude of sex differences in muscle mass may decrease.

Until recently a muscular body was frowned on as unfeminine. Lerner and Karabenick (1974) suggest that self-esteem in girls is related to attractiveness whereas self-esteem in boys is a function of perceived fitness and strength. While an athletic body helps boys attain social prestige and proves their toughness, athleticism in girls has traditionally been associated with tomboyism. When a group of 19 year olds were asked why they had given up being tomboys one commented, ''I began to realize that I was losing my femininity while I was trying to prove my athletic ability. Now I'm still active but less athletic.''

Title IX has expanded athletic opportunities within the schools, helping girls learn physical skills, body control, competitive spirit, and the dynamics of group effort. There is evidence that team members become more independent, creative and autonomous. Young female athletes also learn to view their bodies not merely as objects of decoration, but as sources of strength and pleasure (Richmond-Abbott, 1983). Stigmatization of female athletes continues however, especially in sports which involve body contact or those which are not ''aesthetically pleasing.'' Participants in such sports are considered less feminine to begin with or are thought to become less feminine through participation. Over 50% of female basketball players,

40% of swimmers and 31% of gymnasts felt they were negatively judged (Snyder & Spreitzer, 1978). The media recently has begun to portray muscular models as attractive (see *Vogue,* April 1983). The latest Miss Universe was a physical fitness enthusiast and the Jane Fonda transformation is seen everywhere. This new emphasis on fitness is motivated both by a desire for beauty as well as for health. In fact, the two are often confounded. For example, Lydia Pynkham's famous vegetable compound was first created to cure "female complaints," but later sold for beauty purposes. *The New York Times* runs a combined Health/Beauty column featuring a kind of new enlightened narcissism based on health factors. Exercise is being sold to girls as nature's best makeup, and big profits are made from chic accessories which help turn girls into fashionable jocks.

The current fusion of fitness with beauty images has, for some young women, increased anxiety about their bodies. Girls used to say "I don't want ugly muscles" whereas now they say "I'm ugly because I have no muscles." Thin = beautiful is being replaced by strong = beautiful. As women aspire to equality with men, they are also mimicking men's angularity and muscular build. Some adolescent girls are so focused on changing their appearance rather than on feeling good that they become totally obsessed with exercise. In fact, compulsive exercising is a common symptom among anorectics. The incidence of eating disorders is also especially high among gymnasts, skaters and dancers for whom the aesthetic dimension of slimness are essential (Combs, 1982). Adolescent girls should certainly be encouraged to pursue the positive rewards of athletics, but also cautioned that the new "muscular beauty mystique" can lead to commercial exploitation and self-destructive misuse of their bodies. Hutchinson (1982) outlines a therapeutic program which employs fitness training to foster positive body image and self-esteem, and which teaches young women to be less judgmental and more accepting of their natural physique.

Cosmetics and Acne

Cosmetics are a basic ingredient of feminine beauty and, for teenage girls, the use of makeup is one of the important initiation rites into womanhood. Makeup serves as a sexual signal and helps to exaggerate sex differences. A problematic relationship exists between cosmetics and health, for makeup is both an insidious cause of skin problems as well as a useful cover-up. Acne is suffered by most adolescents. It seriously affects one in four. Onset is typically at early puberty, reaching a peak in severity between ages 18-21. Genetic in origin, this disease is triggered by increasing testosterone levels which in turn stimulate the function of sebacious follicles. Females report flare-ups in the third week of the menstrual cycle, probably due to increased progesterone levels (Fulton & Black, 1983). Acne has a greater psychological impact on females than on males because an attractive appearance is more essential for a girl's social acceptance. Boys can more easily define themselves through sports or academics, thus achieving respect despite their skin blemishes. A form of acne found almost exclusively in adolescent girls called acne excorée is caused by compulsive picking at trivial blemishes. These girls gouge their faces as a means of self-punishment to confirm their feelings of worthlessness (Fulton & Black, 1983).

About one third of the girls who regularly use cosmetics can expect blemishes which dermatologists are now calling acne cosmetica. Many serious cases develop as a direct result of daily makeup routine, for cosmetic ingredients are so potent that they can induce a form of acne even in those who are not genetically prone. It may take several months for cosmetic acne to develop, hence the cause usually remains unsuspected and a vicious cycle ensues; the worse the acne becomes, the more makeup is used to conceal it which only escalates the condition even further. Fulton and Black (1983) offer a step-by-step approach for treating acne, including a list of popular cosmetics, each rated for its degree of acne-causing ingredients (see Table I). Makeup foundation, even the so-called oil-free types, are judged the worst offenders; water-based or glycerin-based products are the least harmful. Girls need greater awareness that cosmetics are the cause of many serious cases of acne, and manufacturers should be encouraged to produce a line for adolescents free of harmful ingredients.

As a final point, the effects of birth control pills on acne are mixed. Androgen dominant or so-called mini-pills (Ovral, Lestrin, Norinyl) tend to aggravate acne, but can take as long as two months to cause blemishes and therefore may go unsuspected. Estrogen dominant pills (Enovid-E and Ovulen) have a very positive effect on acne and sometimes miraculously clear up difficult cases (Fulton & Black, 1983).

Table I
Ratings of Cosmetics and Cosmetic Ingredients
and their Affect on Acne

1. Examples of Acceptable Cosmetics:

   Almay Pure Beauty Foundation Lotion for Oily Skin
   Helena Rubinstein Bio-Clear
   Revlon Natural Wonder Oil Free Base

   Cosmetics composed of simple, loose iron oxide powders

2. Examples of Unacceptable Cosmetic Ingredients (acne-causing):

lanolins	isopropyl palmitate
isopropyl myristate	isostearyl neopentanoate
myristyl myristate	decyl oleate
butyl stearate	PPG 2 myristyl propionate
laureth-4	sodium lauryl sulfate
isopropyl isostearate	oleyl alcohol
propylene glycol mono stearate	

Adopted from Fulton and Black (1983).

Cosmetic Breast Surgery

Historically and cross culturally, the search for beauty has often involved body mutilation such as piercing, scarring, stretching and even amputation of human flesh. Sophisticated surgical techniques for remodeling body parts now exist and the majority of cosmetic surgery patients are female. In America alone, an estimated 1.2 million breasts have been enlarged or reduced in the past 25 years (Wagner & Imber, 1979). Girls as young as 14 are now seeking augmentation or reduction of their breasts.

Breast development begins early in the sequential stages of puberty, sometimes by the age of 8. Breast size, like penis size, is popularly used as the single index of sexual development (Unger, 1979). Because breasts are so highly symbolic of femininity in our culture, many physically normal girls experience an almost paralyzing self-consciousness about their breast size. Analysis of magazines directed toward young women indicates that 6% of the advertisements are related to breast "improvement" (Kinzer, 1977). Breast satisfaction is associated with positive self-concept and also related to breast size (Lerner, Karabenick, & Stuart, 1973). There is no relationship, however, between rate of adolescent breast development and later breast satisfaction (Kelly & Menking, 1979). Very large breasted women tend to be more dissatisfied with their breasts than small breasted women. Moreover, women with large breasts are perceived by others to be "unintelligent, incompetent, immoral, and immodest" (Kleinke & Staneski, 1980).

Here is a good example of how beauty stereotypes interact with maturational processes to produce adjustment difficulties. Since girls enter puberty about two years ahead of boys, full breasted girls who develop early must carry the burden of their early maturation "up front." Early puberty makes them socially as well as physically "deviant." They often experience tremendous embarrassment, loss of prestige, as well as overt ridicule. Consequently they may avoid athletic participation, refuse to wear bathing suits, and become socially withdrawn. To such girls, surgical correction of their breast problem has great appeal (and in some extreme cases may be a justifiable consideration).

Formerly, girls had to wait until they were 19 or 20 before reduction mammoplasty, however this operation is now being performed on much younger girls. One plastic surgeon warns that it is the parent who often initiates the request for surgery and he cautions doctors to be certain that the teenager herself really wants the operation, and that she is given enough facts to realistically assess the eventual pain, scarring and swelling. Complications are common and include unsightly scarring and loss of nipple sensitivity.

Considering the cognitive and emotional immaturity of most young adolescents, their hypersensitivity to peer rejection, their vulnerability to impulsive behavior and their difficulties in acquiring a positive body image, is it reasonable to expect them to make a wise decision about such a serious and permanent procedure? Doyle (1982) found that college age women have typically had negative feelings about their breasts during pubertal development, but their feelings had become much more positive by the end of adolescence. Those of us concerned with adolescent health must caution young girls who are now being seduced into surgery which they may later regret.

CONCLUSION

In the process of gender role acquisition, children pass from the first stage of gender confusion in early childhood to the second stage of gender conformity during adolescence (Pleck, 1975). Hence, the adolescent girl strives to adopt a stereotypic feminine image as reflected in prevailing beauty norms. Although cultures change, exaggerated demands for attractiveness continue to influence and undermine her psychological adjustment and physical health. The increasing power of the media to define standards of appearance, coupled with new sophisticated technology, are likely to further escalate the problems discussed in this paper.

Yet more positive counter pressures from the women's movement are providing young girls with alternative models; encouraging them to accept natural differences, to appreciate diversity of appearance, and to cultivate strong, competent bodies rather than merely decorative ones. The third and final stage of gender acquisition is reached when highly polarized sex roles are replaced by more androgynous ones that are flexible and situationally adaptive. Current social change may help a greater number of adult men and women to achieve gender transcendence, thereby redefining the nature of masculinity and femininity. As girls grow up with less polarized gender models, they will be less likely to equate femininity with beauty, and better able to adjust to the changes of adolescence.

REFERENCES

Berscheid, E., Dion, K., Walster, E., & Walster, G.W. (1971). Physical attractiveness and dating choice: A test of the matching hypothesis. *Journal of Experimental Social Psychology. 7*, 173-189.

Bruch, H. (1973). *Eating disorders.* New York: Basic Books.

Canning, H., & Mayer, J. (1966). Obesity: Its possible effect on college acceptance. *New England Journal of Medicine, 275*, 1172-1174.

Clifford, E. (1971). Body satisfaction in adolescence. *Perceptual and Motor Skills, 33*, 119-122.

Combs, M.R. (1982, February). By food possessed. *Women's Sports*, 11-17.

Condry, J.C., & Dyer, S.L. (1977). Behavioral and fantasy measures of fear of success in children. *Child Development, 48*, 1417-1425.

Coombs, C.H., Coombs, L.C., & McClelland, G.H. (1975). Preference scales for number and sex of children. *Population Studies, 29*, 273-298.

Douvan, E., & Adelson, J. (1966). *The adolescent experience.* New York: Wiley.

Doyle, K. (1982). *Breast satisfaction and college women.* Unpublished manuscript. College of New Rochelle, N.Y.

Dwyer, J., Feldman, J.J., & Mayer, J. (1967). Adolescent dieters: Who are they? *American Journal of Clinical Nutrition, 20*, 1045-1056.

Elkind, D. (1978). Understanding the young adolescent. *Adolescence, 13*, 127-134.

Erikson, E. (1968). *Identity: Youth and crises.* New York: Norton.

Faust, M. (1979). Physical growth of adolescent girls; Patterns and sequence. In C.B. Kopp (Ed.), *Becoming female: Perspectives on development* (pp. 427-447). New York: Plenum.

Frazier, A., & Lisonbee, L.K. (1960). Adolescent concerns with physique. *School Review, 58*, 387-405.

Freud, S. (1965). *New introductory lectures on psychoanalysis.* New York: W.W. Norton. (originally published in 1933).

Fulton, J.E., & Black, E. (1983). *Dr. Fulton's step by step program for curing acne.* New York: Harper & Row.

Gilligan, C. (1982). *In a different voice.* Cambridge: Harvard University Press.

Halmi, K.A., Falk, F.R., & Schwartz, E. (1981). Binge eating and vomiting: A survey of a college population. *Psychological Medicine, 11,* 697-706.

Halmi, K.A., Goldberg, S.C., & Cunningham, S. (1977). Perceptual distortion of body image in adolescent girls. *Psychological Medicine, 7,* 253-257.

Hoffman, L.W. (1975). The value of children to parents and the decrease in family size. *Proceedings of the American Philosophical Society, 119,* 430-438.

Hutchinson, M.C. (1982). Transforming body image: Your body, friend or foe? *Women and Therapy, 1*(3), 59-68.

Hyde, J.S., & Rosenberg, B.G. (1980). *The psychology of women* (2nd ed.). Lexington, Mass.: D.C. Heath.

Kelly, H., & Menking, S. (1979). Recalled breast development experiences and young adult breast satisfaction and breast display behavior. *Psychology, 16,* 17-24.

Kinzer, N.S. (1977). *Put down and ripped off: The American woman and the beauty cult.* New York: Cowell.

Kleinke, C., & Staneski, R. (1980). First impressions of female bust size. *Journal of Social Psychology, 10,* 123-124.

Lanier, H.B., & Byrne, J. (1981). How high school students view women: The relationship between perceived attractiveness, occupation, and education. *Sex Roles, 7*(2), 145-148.

Lerner, R.M., & Karabenick, S. (1974). Physical attractiveness, body attitudes and self concept in late adolescence. *Journal of Youth and Adolescence, 3,* 307-316.

Lerner, R.M., Karabenick, S., & Stuart, J.L. (1973). Relations among physical attractiveness, body attitudes, and self-concept in male and female college students. *Journal of Psychology, 85,* 119-29.

Levenkron, S. (1982). *Treating and overcoming anorexia nervosa.* New York: Charles Scribner's Sons.

Lott, B. (1981). *Becoming a woman: The socialization of gender.* Springfield, IL: Charles C. Thomas.

Maccoby, E.E., & Jacklin, C.N. (1974). *The psychology of sex differences.* Stanford, CA: Stanford University Press.

Mayer, A. (1982, July). The gorge purge syndrome. *Health,* pp. 5-52.

Musa, K.E., & Roach, M.E. (1973). Adolescent appearance and self concept. *Adolescence, 8,* 385-394.

Offer, D., Ostrov, E., & Howard, K. (1981). *The adolescent: A psychological self portrait.* New York: Basic Books.

Petersen, A. (1979). *The psychological significance of pubertal changes to adolescent girls.* Paper presented at the Society for Research in Child Development, San Francisco.

Pleck, J.H. (1975). Masculinity-femininity: Current and alternative paradigms. *Sex Roles, 1,* 161-178.

Richmond-Abbott, M. (1983). *Masculine and feminine: Sex roles over the life cycle.* Reading, Mass.: Addison-Wesley.

Rindskopf, K.D., & Gratch, S.E. (1982). Women and exercise: A therapeutic approach. *Women and Therapy, 1*(4), 15-25.

Rivers, C., Barnett, R., & Baruch, G. (1979). *Beyond sugar and spice.* New York: Ballantine Books.

Rubin, J.Z., Provenzano, F.J., & Luria, Z. (1974). The eye of the beholder: Parents views on sex of newborns. *American Journal of Orthopsychiatry, 44,* 512-519.

Snyder, E., & Spreitzer, E. (1978). Correlates of sports participation among adolescent girls. *Research Quarterly, 47,* 804-808.

Stannard, U. (1971). The mask of beauty. In V. Gornick (Ed.), *Woman in sexist society.* New York: Basic Books.

Tan, A.S. (1979). TV beauty ads and role expectations of adolescent female viewers. *Journalism Quarterly, 56,* 283-288.

Umiker-Sebok, J. (1981). The 7 ages of woman: A view from American magazine advertisements. In C. Mayo & N. Henley (Eds.), *Gender and non-verbal behavior.* New York: Springer-Verlag.

Unger, R.K. (1979). Female and male: *Psychological perspectives.* New York: Harper & Row.

Unger, R.K. (1981). *Personal appearance & social control.* Paper presented at the First Interdisciplinary Congress on Women, Haifa, Israel.

Wagner, K.J., & Imber, G. (1979). *Beauty by design.* New York: McGraw-Hill.

Wallston, B.S., & O'Leary, V.E. (1981). Sex and gender make a difference. In L. Wheeler (Ed.), *Review of personality and social psychology: Vol. 2.* Beverly Hills: Sage.

The First Pelvic Examination and Common Gynecological Problems in Adolescent Girls

Karen Hein, MD

ABSTRACT. The first pelvic examination is a rite of passage into American womanhood. Indicators and special considerations for performing a pelvic examination for a teenaged female are reviewed. Factors in the history, physical examination, and laboratory evaluation of the adolescent are presented. The sequence of pubertal events before, during, and after menarche is discussed so that menstrual disorders are presented in the context of normal development. The author presents a synthesis of her approach to caring for adolescents with a review of current literature and resources available to health professionals.

INTRODUCTION

The first pelvic examination is a rite of passage into American womanhood. Like most rituals there has evolved a set of myths and meanings that are associated with it. The experience is never forgotten. Recollections of adult women are filled with emotions revolving around the experience that were not usually shared with family or physician at the time of the first pelvic examination but can be vividly recalled many decades later. Few women recall the experience as a positive one. For many, it raised questions about the health of their bodies rather than reassuring them. For some, it is recalled as a painful or frightening episode with little preparation or explanation before, during or after.

A pelvic examination is often linked with four female life experiences: menarche, marriage, maternity and menopause. In between these major events, adolescents may seek care for specific complaints or needs including contraception, STD, or menstrually related problems. Only recently have we begun to question the nature of the experience of the first pelvic examination. It is my hypothesis that the experience can be a dialogue between patient and health care professional, an opportunity to learn more about one's body, a chance to ask and answer questions about pelvic parts, a time to screen for potential problems and lastly, an opportunity to demystify the internal and external changes that accompany puberty.

Adolescence is the usual time when the first pelvic examination is often performed. It is my belief and experience in caring for hundreds of adolescent females that it need not be unpleasant (Hein, 1981). By attending to the following three emotions, the experience can, indeed, be a positive one for patient and health provider. (1) Fear (of the unknown, of finding pathology, of exposure of genitals). (2) Ignorance (of what is being done, why it is being done, and how it is being done). (3) Discomfort (physical or psychological, that may be compounded by lack of sensitivity or training on the part of the examiner). It should be the role of the health care provider, not the patient, to attend to these barriers.

Two new developments in the past decade aid both patient and health professionals alike in converting the first pelvic examination into a positive experience. The first is the availability of GTA (gynecologic teaching associates) or similar programs (Sarrel, 1981). GTAs are women who receive special training to instruct medical personnel in the proper and appropriate techniques needed for pelvic examination. Through role playing and instruction, GTAs effectively transmit factual and attitudinal information to the trainee. The second advance is the proliferation of excellent visual aids in the form of booklets, pamphlets, and books to guide both the health professional and patient in our understanding of the anatomy, physiology, and functioning of various internal and external organs (Demarest & Sciarra, 1976).

An ever expanding list of types of professionals may be on the team of health professionals involved in the first pelvic examination. The pediatrician, gynecologist, internist, family practitioner or other primary care physician, nurse, nurse-practitioner, mid-wife, physician's assistant or associate, each may touch the life of an adolescent at this time. Knowledge and skill in dealing with adolescents, not the specific medical degree, should be the criterion used to decide who is the most appropriate to perform the first pelvic examination.

INDICATIONS FOR THE FIRST PELVIC EXAM

The completeness of the pelvic examination will depend upon the indication (Emans, 1983). A simple inspection examination of the external genitalia should begin on the first day of life and then be continued routinely at each health maintenance examination throughout childhood. Internal examination is indicated for the specific reasons listed in Table 1. Questions about "virginity" (i.e., penetration beyond the hyman) should not be an indication for performing a pelvic examination.

The examiner cannot reassure the patient that she is developing normally unless the secondary sexual characteristics are seen first. The excuse that the patient would be too embarrassed to have a genital examination usually denotes reluctance on the part of the examiner rather than the patient. In general, the two main reasons to per-

TABLE 1: Indications for the first gynecologic exam

1. Menstrual abnormalities such as amenorrhea,
   menometrorrhagia (or intermenstrual bleeding),
   or dysmenorrhea.

2. Abdominal pain, because of the possibility
   of salpingitis (with or without perihepatitis),
   ovarian cyst, neoplasm, ectopic pregnancy,
   or hematocolpos.

3. Sexual activity, to obtain specimens for
   detection of asymptomatic gonorrhea or cytologic
   abnormalities, and in conjunction with
   prescription of birth control.

4. Vaginal discharge, in order to obtain material
   for "wet prep" and culture to facilitate diagnosis
   and treatment.

5. Maternal history of DES exposure during pregnancy.
   On the basis of this history alone, the patient
   should be referred to a gynecologist for more
   thorough evaluation.

6. The teenager's desire to know that she is "normal."

7. Rape.

From: Hein, K. (1981). The first gynecologic examination. *Diagnosis, 3,* 32-52. Reprinted with permission.

form an internal examination are to answer specific questions that are raised by the physician or by the patient herself.

PROCEEDING THROUGH THE FIRST
PELVIC EXAMINATION

Before the Exam

A verbal contract between the health care provider and patient must be established before the exam is performed. Confidentiality regarding sexual matters should be discussed and assured by the examiner. The general categories in the history and physical examination of the teenager to be covered in the remaining portion of the visit, should be outlined in the beginning. Often it is the first time that new "rules" are being applied to the teenager so that the guidelines should be reviewed first. In all 50 states, young people can be diagnosed and treated for venereal disease without parental consent or knowledge. Since VD can be asymptomatic and therefore detected only by culture of internal sites, performing an internal examination requires the consent of the patient (not the parent or guardian). The choice of having an accompanying adult in the room should be the adolescent's (not the parent or health care provider).

Screening History

The adolescent views herself as a teenager first and not as a gynecologic patient. Asking additional information about other spheres of activities such as friends, family, and interests, gives the feeling that the health care provider sees her as a person. Obtaining this information has medical relevance as well as personal significance. Areas that should be covered and sample questions that should be asked are given on Table 2. Two elements are vital at this stage. First, getting information that is specific and relevant to the patient's level of development and experience. A complete sexual history unbiased by the examiner's personal prejudices or judgments about behaviors is required. Second, information about the youngster's view about her own body as well as her knowledge and views on tampons and her views of previous positive and negative sexual experiences open avenues for later discussion.

The initial interview is the time to introduce visual aids in the

TABLE 2:   Relevant Screening History of an Adolescent

Family history

    What sexual topics are discussed, and how are they handled?

    Have specific complaints or areas of friction arisen between

    the patient and her parents regarding sexual activity?

    How does the family deal with privacy?

DES

    Was DES or any other medication taken during pregnancy?

Peer relationships

    What sort of activities does the girl engage in with peers?

    Does she feel pressured by friends?

    Who helps her decide standards of conduct?

Social role

    Does the youngster go to school or work; or both?

    What community or religious center activity ties does

    she have?

Personal history

    Does the youngster date?

    Have any physical experiences made her uncomfortable?

    Has she kissed, petted, or been touched on the breasts

    or vaginal area?

    Has she had intercourse?

    Does she use tampons? Why or why not?

From: Hein, K. (1981). The first gynecologic examination. *Diagnosis, 3,* 32-52. Reprinted with permission.

form of plastic models, pictures or illustrations. Elements of crucial importance are that the visual aids be (1) accurate, (2) three-dimensional, (3) portray the feminine body in positive light (not showing pathologic conditions or frightening exaggerated colors), (4) show the relationships of anatomic parts to each other and, (5) include male as well as female reproductive organs. The reason to have adequate visual portrayals is that many teenagers cannot yet think abstractly. Therefore, things which they cannot literally see are not well visualized conceptually. Line drawings and sketches are not helpful for most adolescents. Excellent inexpensive texts are now available for use in instructing patients (Demarest & Sciarr, 1976).

There is no reason to accomplish the entire interview, educational process, and complete pelvic examination in one session. Some youngsters require more time to understand and incorporate the concepts and procedures that are required in the pelvic examination. Having all the equipment available to show the adolescent before proceeding with the exam is very important. While seated and dressed in her normal street clothes, the patient should be allowed to hold and inspect the speculum, pap smear equipment, as well as any contraceptives being discussed. During this preliminary discussion, the instructor can transmit certain attitudes about the female body. For example, when explaining the location and size of the cervix one might say, "if you put your finger inside your vagina, at the end you will feel your cervix which feels something like the tip of your nose," thereby transmitting the message that it is permissable to explore one's own body.

During the Exam

After urinating, the patient should be instructed to change from her clothes into a gown in a room without the examiner present. The gown is essential to provide easy visualization by the examiner while limiting extent of exposure of the patient. The decision about additional use of a drape should be up to the teenager. She may wish to see external or internal structures with the aid of a hand mirror. Whether she chooses to watch the examination or not, maintaining eye contact and verbal contact with the patient is necessary and reassuring for all adolescents.

Inspecting the external genitalia. The presence and distribution of pubic hair is noted according to Tanner staging. Examining for the presence of inguinal lymphadenopathy, pubic lice, scabies, sores, etc., is accomplished by simple visual inspection. The size, color and shape of the external genitalia including the labia majora, minora and clitoris as well as the presence or absence of vaginal discharge, perineal skin lesions (e.g., venereal warts, swelling of Skene's and Bartholin's glands) are noted.

Vaginal discharge. Vaginal discharge may be physiologic or indicate the presence of a local vaginal infection (e.g., trichomoniasis, Gardnerella vaginalis), or an infection of the cervix (herpes, chlamydia or gonorrhea), or other higher genital tract infection. Looking at a sample of the discharge mixed with several milliliters of saline under a microscope ("wet prep") can differentiate a physiologic discharge by the presence of multiple epithelial cells with only a few "pus" or polymorphonucleated leukocytes from a pathologic discharge. If true infection is present, the organisms themselves (e.g., fungi such as monilia or parasites such as pin worms or trichomoniasis) are seen. Alternatively, specific cellular changes (e.g., "clue cells" with Gardnerella vaginalis or multinucleated giant cells in the case of herpes) can be seen. Ideally, the specimen for "wet prep" and culture should be collected by direct visualization from the site where the infection occurs. Since some cultures (gonorrhea and chlamydia) should be obtained from within the cervical canal rather than from the vagina, direct visualization of the cervix is important for obtaining certain cultures.

The Internal Examination

Having taken the time to explain the reasons for the internal examination and familiarizing the teenager with the equipment beforehand should make the speculum exam no more uncomfortable or unpleasant than the speculum exam of the ears in a child. There are a variety of types and sizes of specula than can be selected and matched to the patient's preference and size. Disposable plastic specula manufactured in several sizes are preferred by many teenagers. They find the plastic less frightening than the metal. In addition, some models are fitted with the light which eliminates the need for a bulky overhead lamp. The combined unit with a presterilized speculum makes the pelvic examination very portable. It can be performed on any examining table without the need for a special pelvic exam table or even in bed in the case of a hospitalized teenager. Alternative specula include the Hoffman speculum (1/2 × 4-1/2 inches), the Pederson speculum (1 × 4-1/2 inches), the Graves speculum (1-3/8 × 3-3/4 inches).

Bimanual examination is performed in order to feel the size, shape and mobility of internal structures that cannot be visualized. The uterine body, cervix and the adnexae including Fallopian tubes and ovaries are assessed during this part of the pelvic examination.

Rectovaginal palpation is also necessary to assess the septum between the vagina and rectum (especially if the patient was exposed to DES in utero) and in order to palpate the uterus if retroverted or retroflexed. Using the analogy that the rectal exam is like having one's temperature taken with the use of a rectal thermometer helps the patient relate this part of the exam with a familiar past experience.

At each step of the examination explanations are given to the patient about what structures are being palpated and reassurances offered as the exam progresses. After the examination is complete, the examiner should help the patient sit up and then complete the physical examination with a procedure that makes the patient feel less vulnerable and exposed. Examining the spine for scoliosis is my way of ending the physical examination since this requires that the patient be standing rather than lying on the examining table. Patient should then be asked to dress into street clothes in private before discussion or concluding remarks take place.

Post Examination Discussion.

Discussion of the findings should be done with the patient dressed and seated facing the examiner once again, never while patient is lying down being examined. Audio visual aids should be used to explain the location or function of any organs being discussed. Reassurance about normal growth and development can now be offered to balance any specific fears and to form the basis of discussion for the questions raised by the patient.

SEQUENTIAL EVENTS OF PUBERTY

The sequence of events in puberty is usually predictable but the chronological age at which they occur varies greatly. The appearance and changes in the secondary sexual characteristics have been modified by Drs. Tanner and Whitehouse into a sex maturity rating index based on simple inspection examination of breast, pubic hair and the appearance of external genitalia (Figures 1 and 2) (Tanner, 1962). The order of progression is usually that breast budding and the appearance of pubic hair are followed by the growth spurt, which in turn, is followed by menarche usually within two years of breast budding. Menarche, therefore, is a relatively late pubertal event. A summary of pubertal events is given in Figure 3.

COMMON GYNECOLOGICAL DISORDERS

In adolescence the most common gynecologic complaints have to do with the onset, frequency, or duration of periods (Hein & Litt, 1980). No menses (amenorrhea), infrequent menses (oligomenorrhea), too much bleeding at the time of menses (menorrhagia), too

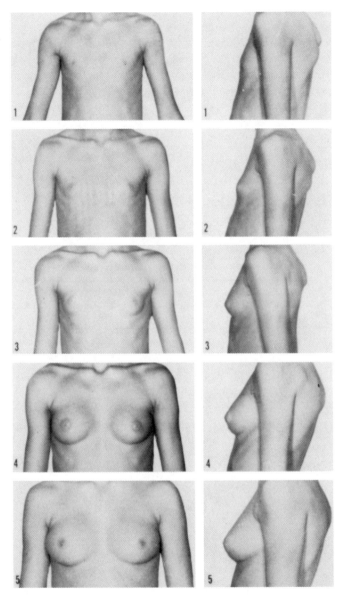

FIGURE 1. Standards for breast development ratings during adolescence. From: Tanner, J. *Growth at adolescence.* Blackwell Scientific, London, 1962, facing p. 37. Reprinted with permission.

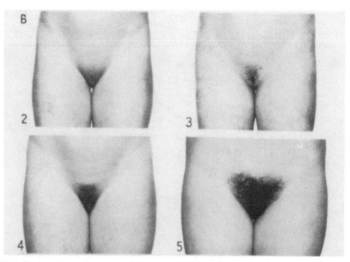

FIGURE 2. Standards for pubic hair ratings in girls. From: Tanner, J. *Growth at adolescence.* Blackwell Scientific, London, 1962, facing p. 33. Reprinted with permission.

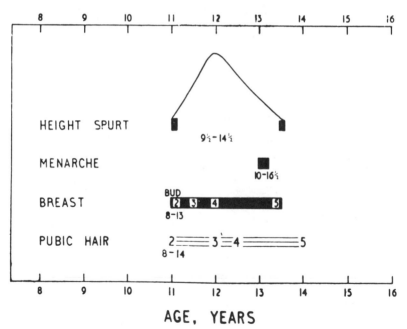

FIGURE 3. Diagram of sequence of events at adolescence in girls. An average girl is represented: the range of ages within which some of the events occur is given by the figures placed directly below them. From: Tanner, J. *Growth at adolescence.* Blackwell Scientific, London, 1962, p. 36. Reprinted with permission.

frequent episodes of bleeding (metrorrhagia), or painful periods (dysmenorrhea) constitute the most frequent complaints. The first pelvic examination may be performed as part of the evaluation of one of the menstrual disorders. Menstrual irregularities are so frequent during adolescence that it behooves all health professionals caring for teenagers to be familiar with the usual spectrum and problems associated with menarche and subsequent menses (Litt, 1983).

Amenorrhea

Amenorrheas may be primary, meaning that menses never began or secondary, meaning that menses have ceased for a period of three or more months. Menarche occurs in this country at an average age of 12.7 years. However, the range of normal is 10-16 years of age. Menarche normally occurs within one year of age of when the patient's mother had her first period. Often the age of menarche of siblings is helpful in trying to arrive at what is normal for a given patient. Menarche usually occurs within two years of the onset of the appearance of breast budding or pubic hair. Low body weight or perhaps more importantly, low body fat associated with eating disorders and exercise, are known to significantly delay the onset of menses and the establishment of regular periods.

The reasons for delayed onset of periods or secondary amenorrhea are multiple. They are summarized in Table 3. Congenital anomalies must be considered as well as chronic illnesses. The range of chronic illnesses associated with amenorrhea is shown in Table 4. Psychiatric illnesses including anorexia nervosa and low weight from exercising are becoming increasingly common reasons for delay in menarche or secondary amenorrhea. Drugs which may be associated with amenorrhea are listed in Table 5. Those causes directly related to gynecologic functions such as ovarian and uterine problems should also be included in the differential diagnosis of amenorrhea.

The evaluation of adolescents with amenorrhea should include a very careful history outlining the previous growth to that point including the timing of development of secondary sexual characteristics, the family pattern of maturation, medications and a careful sexual history. Pregnancy can occur even before the first period if ovulation precedes menarche. A thorough physical examination including notation of anomalies or cyanosis as well as anthropometric records of height and weight, vital signs, careful inspection of the secondary sexual characteristics for assignment of stage of maturity, specific examination of the thyroid, cardiac, skin, and eye grounds are all important in order to eliminate causes listed in the Table above. Pelvic examination including inspection and bimanual exam usually suffice to be sure that the uterus and cervix are present and to detect any masses in the adnexae. Basic laboratory evaluation includes assessment for anemia, infection or inflammation by sedimentation rate, urinalysis, and pregnancy test. If these screening tests are negative, an evaluation of the sella turcica by x-ray or a CT scan, assessment of gonadotropins (LH and FSH), prolactin, thyroid function tests (TSH, T_4) and finally, laparoscopy and ovarian biopsy if indicated based on the above tests. In the coming years with increased availability of LHRH, pinpointing the exact location of anatomic or physiologic dysfunction will become a simpler procedure. Hypothalamic, thalamic, pituitary, or end organ dysfunction will be distinguished from one another.

TABLE 3. Differential Diagnosis of Amenorrhea

Congenital anomalies
    Mullerian duct system
    Gonads (eg, Turner's syndrome)
Chronic illness (see Table 4)
Psychiatric illness
    Anorexia nervosa
    "Boarding school" syndrome
Weight loss
Drugs (see Table 5)
Gynecologic
    Ovarian
        Tumor
        Polycystic ovary
        Cyst
    Uterus
        Pregnancy
        Endometrial hypoplasia

From: Litt, I.F. (1983). Menstrual problems during adolescence. *Pediatrics in Review, 4,* 203-212. Reprinted with permission.

TABLE 4.   Chronic Illness
Associated with Delayed Pubertal
Development and Amenorrhea

Endocrine

   Hypothalamic tumor

   Pituitary

     Insufficienty

     Tumor

   Thyroid

     Hypersecretion

     (Rarely hyposecretion)

   Adrenal

     Insufficiency

     Hypersecretion

   Diabetes mellitus (poorly controlled)

Hematologic

   Sickle cell anemia

   Thalassemia major

Gastrointestinal

   Inflammatory bowel disease

   Cystic fibrosis

Cardiac

   Congenital cyanotic heart disease

From: Litt, I.F. (1983). Menstrual problems during adolescence. *Pediatrics in Review, 4,* 203-212. Reprinted with permission.

Menometrorrhagia

The usual amount of blood loss during the menstrual period is approximately 30 to 60 cubic centimeters (1 to 2 oz). Excessive blood loss is usually frightening to the patient and parents. On occasion, if severe or rapid enough, menometrorrhagia may constitute a medical emergency. Usually excessively frequent bleeding or loss of large amounts of blood are not associated with serious underlying disease if they occur during the first few months after menarche. However, if blood loss leads to severe anemia and loss of blood volume, treatment is indicated to stop the bleeding. Evaluation for an underlying cause may be necessary and appropriate. The patient who is within two years of menarche most commonly has excessive bleeding associated with an anovulatory cycle. The endometrium proliferates

TABLE 5. Drugs that May
Cause Amenorrhea

Hormones

    Estrogens

    Progestogens

    Testosterone

Antihypertensives

    Methyldopa

Antidepressants

    Phenothiazines

Chemotherapeutic agents

    Cyclophosphamide

    Chlorambucil

    Vincaleukoblastine

    Methylhdrazine

Radioisotopes

    ^{32}P

    ^{131}I

    ^{198}AU

Vitamins

    A (in toxic doses)

Illicit drugs

    Heroin

From: Litt, I.F. (1983). Menstrual problems during adolescence. *Pediatrics in Review, 4,* 203-212. Reprinted with permission.

under the stimulation of estrogen but since ovulation does not occur, no progesterone is secreted. Therefore, the endometrium is not converted to secretory tissue and instead, thickens, becomes friable and unstable causing the excessive or prolonged bleeding. Table 6 summarizes other causes of menometrorrhagia.

The treatment for menometrorrhagia should be geared toward (1) correcting the anemia if present, (2) stopping the bleeding if blood loss is so rapid or profound that the patient is symptomatic, (3) treating the underlying condition if anovulatory cycles are initiated by one of the causes on Table 6. In adolescence, treatment is usually medical as opposed to surgical (D&C). Oral administration of an estrogen plus a progestin or estrogen alone or progesterone alone can usually control the blood loss without surgical intervention.

Since cancer is rarely the cause in adolescence, a D&C is rarely needed for diagnostic purposes during the teenage years.

Dysmenorrhea

Painful menses is the most common reason for school absenteeism among adolescent females. Two-thirds of teenagers experience dysmenorrhea and 14% are incapacitated by painful menses. Recent developments which define the role of prostaglandins in the production of menstrual cramps and the subsequent development of therapeutic agents to block the synthesis and action of prostaglandins have revolutionized the approach to the treatment of dysmenorrhea (Hein, 1983). Dysmenorrhea is more common among adolescence as compared to older adults and usually develops after the first few menstrual cycles. The first cycles tend to be anovulatory. Dysmen-

TABLE 6. Differential Diagnosis of Menometrorrhagia

Painless	Painful
Systematic	Trauma
Coagulopathy	Threatened abortion
Congenital	Salpingitis
von Willebrand disease	Intrauterine device
Acquired	
Aspirin sensitivity	
Aplastic anemia	
Anticoagulant treatment	
Neoplasm-bone marrow infiltration	
Idiopathic thrombocytopenia	
Endocrine	
Hypothyroidism	
Oral contraceptives--used improperly	
Local	
Gynecologic	
Dysfunctional uterine bleeding	
Neoplasm	

From: Litt, I.F. (1983). Menstrual problems during adolescence. *Pediatrics in Review, 4,* 203-212. Reprinted with permission.

orrhea is associated with ovulatory menses. New effective agents including fenamates and derivatives of arylproprionic acid are now available to treat dysmenorrhea. Pretreatment before onset of menstruation is not necessary. Prostaglandin production and release is maximal in the first 48 hours of menstrual flow, therefore, these preparations can be prescribed from the onset through the second day of menses. The following are 3 examples of agents that appear to offer comparable relief from dysmenorrhea: Ibuprofen 400 mg 4 times daily; Naproxen 550 mg initially then 275 mg or 250 mg tablets 2 or 3 times per day; Mefenamic Acid 250-500 mg 3-4 times per day.

Since these medications are used for only a few days per cycle, side effects are usually minimal and tolerable. Gastrointestinal symptoms (indigestion or change in stool frequency) are most common but tend to be mild. The long half-life of naproxen (10-17 hours) with corresponding need for infrequent dosages (often twice a day is sufficient) offers an advantage of this preparation for teenagers.

Secondary dysmenorrhea is so common in intrauterine device wearers that routine use of prostaglandin synthetase inhibitors may be justifiable at the time of insertion since their use will reduce cramping as well as the excessive blood loss that is experienced by many young women using an IUD.

Oral contraceptives are also effective in decreasing dysmenorrhea and should be considered for the adolescent also desiring contraceptive protection or for the adolescent with a gastrointestinal ulcer for whom non-steroidal, anti-inflammatory agents are contraindicated. In summary, there are several safe, effective preparations now available to treat dysmenorrhea, a common cause of short term disability among adolescents.

CONCLUSION

In summary, there are special techniques, considerations and approaches that should be used when caring for an adolescent. The first pelvic examination is usually performed during the teenage years. Knowledge of the accompanying physiologic and psychologic events that comprise puberty is crucial in order to place a complaint, a question, or a disorder into the context of the normal sequence of development.

REFERENCES

Demarest, R.J., & Sciarra, J.J. (1976). *Conception, birth and contraception: A visual presentation.* New York: McGraw-Hill.

Emans, S.J. (1983). Pelvic examination of the adolescent patient. *Pediatrics in Review, 4,* 307-312.

Hein, K. (1981). The first gynecologic examination. *Diagnosis, 3,* 32-52.

Hein, K. (1983). Common gynecologic problems. In S. Gellis & B. Kagen (Eds.), *Current pediatric therapy* eleventh edition (in press). Philadelphia: WB Saunders.

Hein, K., & Litt, I.F. (1980). Office management of the adolescent with a gynecologic problem. In A.J. Moss (Ed.), *Pediatrics Update* (pp. 81-102). New York: Elsevier-North Holland.

Litt, I.F. (1983). Menstrual problems during adolesence. *Pediatrics in Review, 4,* 203-212.

Sarrel, P.M. (1981). Indications for a first pelvic examination. *Journal of Adolescent Health Care, 2,* 145-146.

Tanner, J. (1962). *Growth at adolescence.* London: Blackwell Scientific.

The Effect of Pregnancy on Adolescent Growth and Development

Phyllis C. Leppert, MD

ABSTRACT. Adolescent pregnancy represents a triple developmental crisis to young women, that of adolescence, pregnancy, and establishing a relationship with a member of the opposite sex. Physiologically the adjustments of pregnancy are superimposed on those of pubescence. Psychologically and intellectually the pregnant adolescent is still developing, and it is thus difficult for her to meet the physiological demands of pregnancy. Society has responded by developing comprehensive programs for intensive care which have reduced considerably the risks of pregnancy. Contraceptive knowledge has not totally prevented the problems of teenage pregnancy, in part because the adolescents who become pregnant are not sufficiently developed intellectually to perceive that today's actions may prevent future consequences. This paper reviews the literature and synthesizes the published studies with personal experience gained in over five years of professional work with teenage women.

INTRODUCTION

In the decade of the sixties, professionals in health care became increasingly aware that pregnancy in the adolescent years presents developmental challenges to the pregnant teen. In terms of psychosocial development, the gravid adolescent has the tasks of learning to relate to an unborn child and becoming a parent. These tasks are superimposed on the equally difficult tasks of establishing a meaningful bond with a member of the opposite sex and struggling to become her own person intellectually, socially and emotionally. La Barre (La Barre & La Barre, 1969) describes this as a triple crisis. The stress of early parenthood in western society can be devastatingly hard to surmount, in part because financial independence takes a long time to achieve in the developed nations. As a group, women who have their first pregnancies in their teen years have socioeconomic handicaps in later life. Offspring of teenage pregnancies are often handicapped by intellectual, educational and social deficiencies.

Physically, the younger pregnant teenager of sixteen or less is at greater risk of producing a low birth weight baby than her older contemporaries (Coates, 1970). Many studies (Felice, 1981; Perkins, Nakaskima, Mullins, Dubansky & Chin, 1978; McAnarney & Thiede, 1981) conducted in the seventies have shown that the older teenager benefits from comprehensive and intensive antenatal care. It is unclear whether this preventive health service improves the outcome of pregnancy by reducing the incidence of low birth weight babies in teenagers of fifteen and under. Teenagers often do not seek antenatal care because of their inability to understand the concept of prevention and the implications of care for the future (McAnarney, 1980).

No information is available regarding physical growth and development following a pregnancy in very young adolescent girls. Ordinarily, an adolescent would continue to grow in height and to gain weight for approximately two years following menarche, but we do not know if this occurs in young women who become pregnant during their early teens.

In the United States the actual numbers of pregnant teens is rising; however, it is very important to note that the rate of teenage pregnancies is decreasing except in one significant age group—that of 10 to 14 year olds (Blum & Goldhagen, 1981). Adolescent sexual activity is increasing and it is reported that by age 17 about half of the males and one third of the females have had sexual intercourse. In 1978, 1,100,000 adolescent women were pregnant in the United States. In that year 554,000 infants were born to teenagers and 1.3 million children lived with mothers who were teens (Lincoln, 1981). The startling facts are that 20% of pregnancies among adolescent women occur in the first months after initiation of sexual intercourse and about 50% of the pregnancies begin in the first six months of sexual activity (Akpom, Akpom & Davis, 1976; Klein, 1978). Studies show that the majority of these pregnancies are unintended. Eighty-six per cent of the 749,000 pregnancies among unwed adolescents and 51% of the 394,000 pregnancies among married teenage women in 1978 were not planned. Repeat pregnancies during adolescence occurred about 22% of the time as 78% of the births to teenage women were first births (Lincoln, 1981; Currie, Jekel & Klerman, 1972). Women who begin childbearing in the their teens have more children and will have them closer together than women who begin their families in their twenties (Baldwin & Cain, 1980). Studies indicated that the majority of teenagers and in some instances 9 out of 10 young women have already had sexual intercourse before coming to family planning clinics (Zelnick & Kantner, 1978). However, teenage women who desired to avoid pregnancies were able to do so. The overall pregnancy rate among teenagers has risen, but it *has* declined among those who are consistently sexually active. The older the teenage woman who is sexually active, the more she is apt to effectively use contraception. One estimate is that contraception prevents 680,000 premarital pregnancies in teenagers per year (Lincoln, 1981). The number of abortions among teenage women has had the steepest rise, especially among those aged 10-15. In 1978, 378,500 abortions were performed on teenagers (Lincoln, 1981).

The economic impact to our society is great. In 1976, $4.65 billion was paid to households containing young women who delivered their first child while still teenagers; 36% of these women were not high school graduates (Moore, 1978). Teenage parents obtain less education than their contemporaries, and are more often limited to the less prestigious jobs—which for women are usually dead-end ones. The marriages of teen parents are less stable than those who postpone childbearing. These facts persisted throughout their twenties in a carefully matched study of women who delivered children in their teens and a control cohort of their contemporaries. Thus future economic gains to both society and to the pregnant teens themselves are irrevocably lost (Card & Wise, 1978).

To understand the effect of pregnancy on adolescent growth and development, it is imperative first to understand the physiological, psychosocial, and intellectual development of the teenager. We can then examine the superimposed physiology of pregnancy, and the concomitant psychosocial task of initiating a meaningful relationship with the opposite sex and of beginning parenthood.

PHYSIOLOGICAL CHANGES DURING PUBERTY

While adolescence is a psychosocial concept defined in terms of societal norms, pubescence is a physiological phenomenon which unfolds slowly over several years and is a time of transition from a

child's body to that of an adult. Like many transitions it is not necessarily smooth and predictable, but there are nevertheless certain se-
quences of events which occur in a more or less predictable and
orderly manner. The events are under hormonal control, and follow
those hormonal events of childhood which produced longitudinal
growth and skeletal development along with changes in body proportions. In childhood, skeletal proportions, body muscle mass, and
body fluid distribution are different than in adulthood. These facts
affect physiological parameters. For instance, blood pressure is normally lower in children than in adults. The child and younger adolescent's ability to metabolize drugs is not the same as that of adults
and one measurement of red blood cells used to assess anemia, mean
corpuscular volume, is different in childhood. Thus, puberty is a
time for changes in many physiological systems in addition to the
changes in the reproductive organs (Nelson, 1978).

The longitudinal growth of children is controlled by the pituitary
growth hormone, which is secreted in a rhythmical pattern in childhood. Infants and children grow at an extremely fast rate. Then the
velocity slows until the onset of puberty. Pubescent growth occurs
in three phases. In the first phase the rate is minimal and similar to
that of the childhood years. The second period is the most rapid
period of growth and is followed by a time of decreased velocity or
slower growth, which culminates in the cessation of growth and closure of the epiphyses of the long bones of the body. In young girls
the first stage of growth in height and weight begins at age 11 years
on the average. The period of rapid growth starts around age 12 and
is completed shortly before the 14th year. Menarche, which in the
United States occurs at about age 12, begins after the period of rapid
growth. Following menarche, the rapid phase of growth ends. However, growth does continue in a somewhat limited and decelerated
fashion. This stage is completed at age 15 to 16 years.

In 1969, Marshall and Tanner noted the progression of pubertal
development of 192 English girls and described stages of physical
development that are classic. These stages are clinically very useful
for the physician in evaluating whether a teenager's development is
proceeding normally. Breast development correlates with the onset
of the growth spurt and predates pubic hair development. However,
once pubic hair growth begins, development proceeds quickly
through all the stages. In Tanner's original study the mean period
for the breast development stage was 4.2 years while that of pubic
hair took 2.7 years.

These events are under hormonal control. In early puberty (between eight and twelve years of age) the adrenal glands grow considerably, and adrenal hormonal secretion begins its rise. The hormone cortisol will not peak until age 20 but begins to rise in both sexes at puberty. Adrenal androgens are produced primarily as dehydroisoandrosterone. This increase in adrenal hormones is called adrenarche and results in the development of pubic and axillary hair as well as muscle strength.

The physiologically functional thyroid hormones, free T_3 and T_4, have a minimal but distinct elevation during puberty which correlates with the increase in basal metabolic rate. Bone and calcium metabolism is altered during adolescence. PTH (parathyroid hormone) raises blood calcium and CT (calcitonin) lowers it. Together, along with Vitamin D from sunlight, they regulated the metabolism of calcium. From age 6 to 13 years CT is decreased and PTH increased, compared to adult levels, which demonstrates active bone metabolism. Estrogens modify these changes and when they increase in the blood during puberty they help in the closure of the epiphyses.

The developmental changes which produce the events leading to normal menstrual cycles are complex and gradual and are determined by the interplay of the ovarian hormones estrogen and progesterone, the pituitary hormones FSH (follicular stimulating hormone) and LH (luteinizing hormone), and the hypothalamic secretion of G_n RH (gonadotropin releasing hormone).

In childhood, the ovaries are producing low levels of estrogen and progesterone, but because the hypothalamus is then very sensitive to small amounts of ovarian steroidal hormones, the release of G_n RH is restrained and prevents the release of LH and FSH from the pituitary glands. As puberty develops, this situation changes and the hypothalamus becomes less sensitive and allows for the release of increasing amounts of G_n RH. This in turn stimulates the production of more LH and FSH. At first this is produced by the pituitary in a pulsatile pattern. Over time the adult manner of cyclical release occurs with an increase of estrogen at first, stimulating a midcycle peak of LH and FSH. Ovulation occurs followed by the production of progesterone by the ovaries. The endometrial lining of the uterus undergoes changes in preparation for an eventual pregnancy. Classically, the menstrual cycles of adolescents are anovulatory with only an occasional fertile cycle for the first year after menarche. These cycles of bleeding are due to estrogen-induced bleeding at first, and then gradually develop into the full endocrine cycle which includes ovulation on a monthly basis, as seen in adult women. It may take from three to seven years for the complete endocrine cycle to mature, with more ovulatory cycles occurring each year in gradual fashion. Despite this fact, a study of pregnant teenagers in a New York hospital showed that 15% of the girls conceived three to nine months after their menarche (Leppert, 1983).

PSYCHOLOGICAL DEVELOPMENT

Society in developed countries of the world in the 1980s is highly technologically oriented and is apt to be ever more so in the 1990s. It requires the ability to think and reason in very abstract ways. Puberty occurs at a young age and at a time when intellectual and cognitive development is still occurring. To function as an adult economically in a computerized age demands a complex intellectual apparatus that must be educated and well developed. Piaget has emphasized that the cognitive shift or intellectual development that must occur in adolescence is the development of the ability to reason in the abstract. This is the capacity to extrapolate or draw on old experience to solve new problems. Its development is fostered by experience.

Humans only develop this ability by a series of life events which can be thought of as experimentation. From the perspective of intellectual development experimentation becomes a substrate for cognitive growth. The more restricted a person's range of experience, the fewer resources that person has to draw upon for problem solving. The more a person ventures, explores and learns, the more one develops facts to work with and reason about. One psychological device which allows people in general, and teenagers in particular, to sample a broad aspect of life is a sense of invulnerability.

Many of the stresses and strains of adolescence are due to this need to experiment and to try out new ideas and values. Sometimes the stresses are more difficult for the parents of teenagers than for the teenagers themselves. Since early adolescents cannot yet think abstractly, they cannot conceptualize the future. Society needs to set limits, as it does with small children, to protect younger teens from harm that might come from their inability to reason and function completely in adult society. At the same time it is absolutely essential that we promote and allow for productive experimentation among teenagers so that cognitive growth and development can occur.

As an adolescent begins to mature intellectually and emotionally, she develops her own set of values and her own image of herself that does not rely on the opinion of others. She develops an internal locus of control, meaning that she as an individual understands that she has the power to control reactions to life's events and that her actions can make a definite difference in the outcome of those events. A small child is rooted in the present, and has no concept of time. Older children think concretely. They understand what they see, touch and feel. Since they can see pictures of persons such as ancestors who lived before, they can begin to understand history. But the past is concrete. Only the future is abstract.

The ability to reason abstractly as a formal operation facilitates the mastering of the developmental tasks of adolescence: (1) the development of self-identity: who am I?, (2) The determination of one's sexual identification, (3) The attainment of independence and separation from parents emotionally but not rejecting them, (4) The development of a moral value system, (5) A choice of vocation and a commitment to work, (6) The development of the capacity for lasting relationships and for tender gentle love.

Adolescent developmental processes are influenced by psychological and cultural forces as well as intellectual. Early research on teenagers in our society focused on the young people who were disturbed, and overemphasized the stormy stressful aspects of the psychological and cultural adjustments of adolescence. This period is also an opportunity and a time for growth, and many young people master the developmental tasks very well.

PHYSIOLOGICAL CHANGES DURING PREGNANCY

Pregnancy is a time of tremendous physiological adjustment for the woman. The uterus must grow and add new smooth muscle to its walls to accommodate the growing fetus. Breasts enlarge and their ductal system develops as well. During pregnancy a woman also stores fat for energy which will be expanded during labor, delivery and lactation. Anywhere from one half to two-thirds of the weight gained during pregnancy is gained due to this growth and development (Pritchard & MacDonald, 1980).

The endocrine changes of pregnancy are extensive and contribute to altered metabolism. The placenta produces HCG (chorionic gonadotropin) and HPL (placenta lactogen) as well as progesterone and estrogens and a small amount of thyroid—stimulating hormone. Placental lactogen is detectable as early as three weeks after fertilization and is responsible for a large percentage of the altered metabolism of pregnancy. Its concentration rises during the first and second trimester of pregnancy as the placenta grows. This hormone causes breakdown of stored fat, or lipolysis. The end result is an increase in the fatty acids freely circulating in the blood, which provides a source of energy for both the mother and the fetus. The hormone also alters the utilization of glucose in the mother by blocking the uptake of glucose in her cells and by inhibiting the breaking down of storage glucose, allowing for conservation of this carbohydrate. It regulates the maternal production of insulin in such a way that protein synthesis is enhanced, and thus allows for accumulation of amino acids which cross the placenta for fetal needs.

The placenta itself synthesizes estrogen, especially estriol. As early as the seventh week of pregnancy the placenta is responsible for more than half of the estrogens circulating in the mother's blood. The placenta also produces large amounts of progesterone, which allows the uterus to be relaxed during pregnancy and promotes breast development.

These hormonal changes affect the entire maternal organism. The pituitary gland enlarges during pregnancy. Pituitary growth hormone decreases but prolactin increases, especially prior to labor and delivery. Prolactin's exact function during pregnancy is not yet known.

The thyroid gland enlarges and more T_3 (thyroxine) is secreted. However, plasma proteins which bind thyroxine are also increased, so that the amount of free circulating thyroxine which is physiologically active is stable.

In pregnancy PTH (parathyroid hormone) is increased but so is CT (calcitonin). The amount of calcium in the blood stream does not change in pregnancy compared to the non-pregnant state, so that CT may be important to prevent skeletal calcium from being removed from the bones. Cortisol is increased in concentration but it is bound to increased amounts of binding protein.

Ovulation, of course, stops during pregnancy; however, the ovaries begin to secrete a hormone called relaxin. This hormone remodels the connective tissue of the body, making it softer. It is thought to allow for expansion of the pelvic skeleton by changing the composition of the collagen and elastin in the pelvic ligaments. Pregnancy alters water metabolism in that water is retained in the tissues. This phenomenon is characteristic of all pregnancies. An exaggeration of this response produces edema which is usually associated with pre-eclampsia, one of the major complications of pregnancy. Blood volume increases by forty-five percent in the average pregnant woman. This is brought about by the enlarged vascular system supplying the enlarged uterus. A pregnant woman's body also must increase the numbers of circulating erythrocytes to provide for the increase in the blood supply. In order to produce this large number of red blood cells, iron is utilized. It is calculated that 1 gram of iron is needed in a normal pregnancy in an adult woman to provide for the needed increase in erythrocytes.

Pregnancy exerts demands of a great magnitude on the cardiovascular system. Usually the pulse rate increases by 10 to 15 beats per minute. The arterial blood pressure and vascular resistance decrease in normal pregnancies. Since blood volume, maternal basal metabolic rate, and maternal weight increase, the cardiac output is altered. Maternal cardiac output rises continually until 32 to 34 weeks of pregnancy, at which time it decreases.

The increasing size of the pregnant uterus compresses the inferior vena cava and can cause decreased venous return to the heart. It has been clearly shown that cardiac output increases appreciably when a pregnant woman assumes a lateral recumbent position.

The effects of pregnancy occur also in the digestive system, primarily as a decreased gastric emptying and intestinal transit time, and in the urinary tract, primarily as a dilatation of the ureters.

Thus, pregnancy produces the most numerous and intense physiological changes which occur in post natal life. When superimposed on the normal physiological changes of puberty one can see the considerable demands that are made on a pregnant teenage girl.

PSYCHOLOGICAL DEMANDS OF PREGNANCY

Pregnancy can be thought of psychologically as a prelude to parenthood, with all its attendant joys as well as responsibilities. It is a time when all pregnant women tend to become inwardly directed and to focus on themselves and their personal concerns. In very early pregnancy, even before the woman feels fetal movement, she begins to develop an affectionate bond with the unborn fetus. Her unborn baby is a potential child with a unique appearance to her. She fantasizes early in the pregnancy what the child will be like and what he or she will do. She dreams about what the new baby will mean to her and often becomes preoccupied with herself and her thoughts.

Pregnancy is also a time of facing the questions of identity—Who am I? Who will I become? It is a time when all women reassess their situation in life and begin to focus on choices. In a younger woman the choices often have to do with whether or not to continue her education.

Society reacts to the pregnant woman differently and she may begin to feel ugly, fat and unattractive. All of these physical signs emphasize to the pregnant woman that her status is changing, she is about to become a mother with all the overwhelming responsibilities that this implies. She finds her emotions heightened, perhaps because of the extreme hormonal and metabolic changes occurring in her body. She laughs and cries easily and needs a great deal of love and reassurance. Her partner may not understand these changes or know how to respond to them and thus may inadvertently contribute to strains in their relationship. The woman may be quite frightened of the prospect of the new role as a parent. She may repress and deny her fear or may express it as heightened anxiety about parturition or the well being of the unborn child.

Because pregnancy lasts nine months, it allows time for the pregnant woman to begin to adjust and grow into her role as a mother. She learns first to assume responsibility for herself and her own health. During pregnancy, as she cares for her own physical well-being, she promotes the welfare of her baby. Thus maternal responsibility emerges as the mother recognizes that she and her offspring are inextricably linked. With the help and guidance of others she prepares herself for her changed status as a mother.

This developmental task is compounded by her need to establish a relationship with a member of the opposite sex in bonding at a more mature level—a level in which they are parents as well as lovers. That this task is extremely difficult is demonstrated by the high divorce rate in the United States.

When we consider that physiological growth during puberty is juxtaposed with the altered but normal physiology of pregnancy, it is no wonder that pregnancies in the teen years often do not have optimal outcomes. The younger the teenager, the poorer the outcome. When one considers the psychological tasks encompassing both adolescence and pregnancy it becomes apparent that these developmental factors also contribute to the outcomes of the pregnancy.

COMPREHENSIVE PRENATAL CARE

It has been shown repeatedly that comprehensive prenatal care greatly improves pregnancy outcomes and promotes the delivery of healthy, happy babies. However, this challenge is to encourage young girls to come for prenatal care early in the pregnancy. In most studies of pregnancy outcome, teenagers have been shown to produce smaller babies (both preterm and small for dates babies) than older women and to have babies with a higher incidence of congenital malformations, neurological deficits and higher perinatal mortality. In early studies from the 1960s, higher incidences of preeclampsia, anemia, cephalopelvic disproportion and prolonged labors with increase incidence of cesarean section were reported for adolescents (Sarrell & Davis 1966; McGanity, Lettel, Fogelman, Jennings, Calhoun & Dawson, 1969). Subsequent studies have noted that most significant medical risks are for those pregnant adolescents 16 years of age and under (McAnarney & Thiede, 1981; Perkins, Nakaskima, Mullins, Dubansky & Chin, 1978). All studies clearly show that comprehensive prenatal care improves these outcomes.

Teenagers who become pregnant may not seek prenatal care because they may not have developed intellectually and emotionally to the point where they understand that their actions today affect the future. Those of us who work with teenagers know that those who become pregnant are a different group of adolescents from those who are sexually active but contraceptors. Often pregnancy among teenagers is unintended because of this inability to understand the future. In some cases the future is intellectually understood, especially by older adolescents. They may perceive that having a baby will not really affect their future, so why not have a baby? This phenomenon may explain why adolescents in poverty groups have children. They see no way out of the cycle of poverty. Some authors state that the threat of nuclear war may affect teenagers in this manner also.

The impulsiveness and sense of invulnerability also contributes today to the inability of some pregnant teenagers to attend prenatal clinics early in their gestations. The lack of a future sense makes it difficult to encourage teenagers to follow a nutritious diet that will meet their needs and the needs of the unborn fetus. Nutritional education and supplemental foods, especially protein, are helpful in this age group. Evidence suggests that the pregnant teenager, especially if under age 16 or within a few months of menarche, needs to gain more weight on the average than the adult pregnant woman (Naeye, 1981).

Because the teenager is impulsive and unable intellectually to assume as much responsibility for herself and her well being during pregnancy as an older pregnant woman, the comprehensive pregnancy program must meet her where she is and help her to grow and to increasingly assume responsibility. Thus, such programs have walk-in appointment systems, an access system which allows a teenager to register for care and to receive that care prior to demands for financial reimbursement. Evening or late afternoon hours, clinics in schools, or offices convenient to subway and bus transportation are often part of the plans of such programs. Careful follow-up necessitates a system of home visits or phone calls. Educational needs are met both in group "rap sessions" or in individual counselling sessions. The fathers of the unborn baby and the mothers of the young girls (themselves often mothers in their own teenage years) are encouraged to come to the clinic visits.

Psychologically, the young girls must be helped to develop the important affectionate bond with their unborn baby. This is critical because most studies note that the offspring of teenagers are handicapped intellectually. One study (Hardy, Welcher, Stanley & Dallas, 1978) demonstrated that children born to mothers initially 16 years of age or under had a lower IQ as measured by their Stanford-Binet performance. Another study found that 47% of teen mothers showed a limited ability to cope with the stress of parenthood, compared to 26% of older mothers (Furstenberg & Crawford, 1978). The offspring of teenagers were found deficient in eye contact, verbal interactions, physical contact and smiling. As these children who are born to adolescent mothers grow, they are likely to exhibit deficits in cognitive development and are likely themselves to become teenage parents. The process by which a young pregnant girl can begin to help develop parenting skills can be initiated by having her talk about her unborn baby, naming him or her early, and listening to the fetal heart. Almost all of the teenagers found an ultrasound examination, especially real-time ultrasound, to be extremely meaningful, and sensitive staff can utilize this as an opportunity to help initiate the affectionate bonding with the fetus.

Comprehensive prenatal care, however, implies an intensely dedicated and hard working staff who understand exceedingly well the physiology of pregnancy as well as adolescence and the psychological adjustments of both life periods. The staff must be prepared to "mother the mother-to-be" in the best sense of the word. This fact alone may explain why certified nurse-midwives have contributed so much to the development and success of adolescent programs for pregnant teenagers.

SUMMARY

Pregnant teenagers have the physiological and physiosocial demands of pregnancy superimposed on the developmental changes of puberty. Comprehensive programs for intensive prenatal care can ensure improved outcomes for adolescent pregnant girls and reduce the risks of childbearing.

REFERENCES

Akpom, C.A., Akpom, K.L., & Davis, M. (1976). Prior sexual behavior of teenagers attending rap sessions for the first time. *Family Planning Perspectives, 8,* 203-206.

Baldwin, E., & Cain, U.S. (1980). The children of teenage parents. *Family Planning Perspectives, 12,* 34-43.

Block, H. (1981). Adolescence: A problem of society. Editorial. *Journal of Adolescent Health Care, 2,* 143-144.

Block, R.W., Saltzman, S.T., & Block, S. (1981). Teenage pregnancy. *Advances in Pediatrics, 28,* 75-97.

Blum, R.W., & Goldhagen, J. (1981). Teenage pregnancy in perspective. *Clinical Pediatrics 5,* 335-340.

Bongiovanni, A.M. (1983). *Adolescent gynecology: A guide for clinicians.* New York: Plenum Press.

Card, J.J., & Wise, L.L. (1978). Teenage mothers and teenage fathers: the impact of early childbearing on the parent's personal and professional lives. *Family Planning Perspectives, 10,* 199-205.

Coates, J.B. (1970). Obstetrics in the very young adolescent. *American Journal of Obstetrics and Gynecology, 108,* 68-72.

Currie, J.B., Jekel, J.F., & Klerman, L.V. (1972). Subsequent pregnancies among teenage mothers enrolled in a special program. *American Journal of Public Health, 62,* 1606-1611.

Dyrenfurth, J. (1983). Endocrine aspects of menarche and menopause—milestones in the woman's life cycle. In S. Golub (Ed.), *Menarche* (pp. 21-46). Lexington MA: Lexington Books.

Edwards, L.E., Steinman, M.E., Arnold, K.A., & Haravson, E.V. (1980). Adolescent pregnancy prevention services in high school clinics. *Family Planning Perspectives, 12*, 6-14.

Felice, M.E. (1981). The young pregnant teenager. *Journal of Adolescent Health Care, 1*, 193-197.

Field, T., Widmayer, S., Greenberg, R., & Stoller, S. (1982). Effects of parent training on teenage mothers. *Pediatrics, 69*, 705-707.

Finkelstein, J.W., Finkelstein, J.A., Christi, M., Rodi, M., & Shelton, C. (1982). Teenage pregnancy and parenthood. *Journal of Adolescent Health Care, 3*, 1-7.

Furstenberg, C.F., & Crawford, A.C. (1978). Helping teenage mothers to cope. *Family Planning Perspectives, 10*, 322-333.

Hardy, J.B., Welcher, D.W., Stanley, J., & Dallas, J.R. (1978). Long-range outcome of adolescent pregnancy. *Clinical Obstetrics and Gynecology, 21*, 1215-1232.

Hutchins, F.L., Kendall, N., & Rubin, J. (1979). Experience with teenage pregnancy. *Obstetrics and Gynecology, 54*, 1-5.

Klein, L. (1978). Antecedents of teenage pregnancy. *Clinical Obstetrics and Gynecology, 21*, 1151-1161.

Klein, L. (1974). Early teenage pregnancy, contraception, and repeat pregnancy. *American Journal of Obstetrics and Gynecology, 120*, 249-256.

Kreipe, A., Klaus, J.R., & McAnarney, E.R. (1981). Early adolescent childbearing. *Journal of Adolescent Health Care, 2*, 127-131.

La Barre, M., & La Barre, W. (1969). The triple crisis: Adolescence, early marriage and parenthood in the double jeopardy. *The triple crisis. Illigitimacy today.* New York: National Council on Illegitimacy.

Leppert, P.C. (1983). Menarche and adolescent pregnancy. In S. Golub (Ed.), *Menarche.* pp. 195-201 Lexington, MA: Lexington Books.

Levine, S.V. (1981). The anxieties of adolescents. *Journal of Adolescent Health Care, 2*, 133-137.

Lincoln, A. (1981). *Teenage pregnancy. The Problem that has not gone away.* New York: The Alan Guttenmacher Institute.

McAnarney, E.R., Klaus, J.R., Adams, R.N., Tatelbaum, R.L., Kash, C., Coulter, M., Plum, M., and Chainey, E. (1978). Obstetric neonatal and psychosocial outcome of pregnant adolescent. *Pediatrics, 61*, 91-208.

McAnarney, E.R., & Thiede, H.A. (1981). Adolescent pregnancy and chlidbearing: What we learned in a decade and what remains to be learned. *Seminars in Perinatology, 5*, 91-103.

McAnarney, E.R. (1980). Pregnancy in adolescence. *Pediatric Annals Spl. Issue, 9*, 11-75.

McGanity, W.J., Lettel, H.M., Fogelman, A., Jennings, L., Calhoun, E., & Dawson, E.B. (1969). Pregnancy in the adolescent. I. Preliminary summary of health status. *American Journal of Obstetrics and Gynecology, 103*, 773-788.

Moore, K.A. (1978). Teenage childbirth and welfare dependency. *Family Planning Perspectives, 10*, 233-235.

Moore, K.A., & Hofferth, S.E. (1977). *The consequences of early childbearing.* Washington, D.C.: Urban Institute.

Naeye, R.L. (1981). Teenaged and preteenaged pregnancies. Consequences of the fetal-maternal competition for nutrients. *Pediatrics, 67*, 146-150.

Nelson, R.M. (1978). Physiological correlates of puberty. *Clinical Obstetrics and Gynecology, 2*, 1135-1151.

Osofsky, J.M., & Osofsky, H.J. (1978). Teenage pregnancy: Psychosocial consideration. *Clinical Obstetrics and Gynecology, 21*, 1161-1175.

Perkins, R.P., Nakaskima, I.I., Mullins, M., Dubansky, L.S., & Chin, M.L. (1978). Intensive care in adolescent pregnancy. *Obstetrics and Gynecology, 52*, 179-188.

Piaget, J., & Bärbe, I. (1969). *The psychology of the child.* New York: Basic Books.

Pool, C.J., Smith, M.S., & Hoffman, M.A. (1982). Mothers of adolescent mothers. *Journal of Adolescent Health Care, 3*, 41-43.

Pritchard, J., & MacDonald, P. (1980). *Williams Obstetrics 16th ed.*, New York: Appleton-Century-Crofts.

Rauch, J.L., Johnson, L.B., & Burket, R.L. (1971). The management of adolescent pregnancy and prevention of repeat pregnancies. *Health Services and Mental Health Administration Health Reports, 86,* 66-73.

Reycroft, D., & Kessler, A.K. (1980). Teenage pregnancy solutions are evolving. *New England Journal of Medicine, 303,* 516-518.

Ryan, G.M., & Schneider, J.M. (1978). Teenage obstetrics complications. *Clinical Obstetrics and Gynecology, 21,* 1191-1199.

Sarrell, P.M., & Davis, C. (1966). The young unwed primipara. *American Journal of Obstetrics and Gynecology, 95,* 722-725.

Settlage, D.S.F., Baroff, S., & Cooper, D. (1973). Sexual experiences of younger teenage girls seeking contraceptive assistance for the first time. *Family Planning Perspectives, 5,* 223-226.

Teberg, A., Howell, W., & Wingert, W.A. (1983) Attachment and interaction, behavior between young teenage mothers and their infants. *Journal of Adolescent Health Care, 4,* 61-66.

Tietz, C. (1978). Teenage pregnancies: Looking ahead to 1984. *Family Planning Perspectives, 10,* 208-209.

Trussel, J., & Menkin, J. (1978). Early childbearing and subsequent fertility. *Family Planning Perspectives, 10,* 209-218.

Tyrer, L.B., & Josimovich, J. (1977). Contraception in teenagers. *Clinical Obstetrics and Gynecology, 20,* 651-663.

Tyrer, L.B., Mayer, R.G., & Bradshaw, L.E. (1978). Meeting the special needs of pregnant teenagers. *Clinical Obstetrics and Gynecology, 21,* 1197-1215.

Wyshak, G., & Frisch, R.E. (1982). Evidence for a secular trend in age of menarche. *New England Journal of Medicine, 306,* 1033-1035.

Youngs, D.D., Niebling, J.R., Blake, D., Shipp, D.A., Stanley, J., & King, T. (1977). Experience with an adolescent pregnancy program. *Obstetrics and Gynecology, 50,* 212-216.

Zelnick, M., & Kantner, J.F. (1978). Contraceptive patterns and premarital pregnancy among women aged 15-19 in 1976. *Family Planning Perspectives, 10,* 135-142.

Scoliosis:
Diagnosis and Current Treatment

John B. Emans, MD

ABSTRACT. The most common form of scoliosis (lateral curvature of the spine) is adolescent idiopathic scoliosis (AIS). The prevalence of AIS is 10-20 individuals per 1,000 population screened. Progressive AIS affects females more frequently than males. Only 1-2 per 1,000 individuals have progressive AIS and need brace or surgical treatment. Early screening programs for scoliosis and other spinal deformities are instituted in schools in the fifth to ninth grades to achieve early detection of progressive AIS. If detected early, progressive AIS can in most cases be successfully treated by use of an external spinal brace. More severe scoliosis is successfully treated with surgical instrumentation and spinal fusion. Although much progress has been made in understanding the natural history of AIS, its cause remains unknown.

INTRODUCTION

Women are affected by significant scoliosis in a ratio of at least five to one over males. Scoliosis and other spinal deformities have received increased attention from health professionals and the public in the last two decades. The advent of better spinal bracing technology for the control of progressive AIS and improved surgical treatments have provided a strong incentive for early identification of spinal deformities. Mandatory school screening for scoliosis is now prevalent and identifies early many patients with treatable scoliosis. Attention in the lay press and media to newer scoliosis techniques and parental awareness of school screening programs makes this review timely. It is the intent of this article to put in perspective the available information and data for allied health professionals and interested individuals.

A glossary of scoliosis terms is included at the end of this article.

The Normal Spine

The normal spine is composed of 7 cervical, 12 thoracic, 5 lumbar and several sacral vertebrae held by the intervertebral discs, ligaments, and a variety of spinal muscles. The individual bony vertebra vary in shape throughout the different regions of the spine but have the same general features seen in Figure 1. Viewed from the side the normal spine is not perfectly straight but exhibits a normal cervical lordosis, thoracic kyphosis and lumbo sacral lordosis (Figure 2). These spinal contours vary somewhat between individuals and races and with postural habits. Viewed from behind the normal spine is nearly perfectly straight with minor deviations common. Normal spinal contours are controlled by heredity, growth, posture, muscular balance and the central nervous system.

Spinal Deformities

Exaggerations or distortions of the normal spinal contours occur. Thoracic hyperkyphosis (roundback) may be due to postural habits or may reflect an underlying developmental abnormality such as Scheuermann's hyperkyphosis. Lumbar hyperlordosis or "swayback" may be hereditary or postural in origin. Scoliosis or lateral

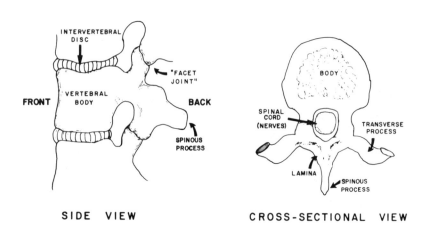

SIDE VIEW   CROSS-SECTIONAL VIEW

FIGURE I
A typical spinal vertebra.

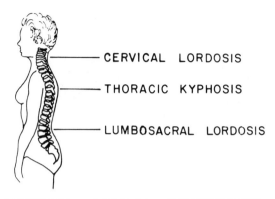

NORMAL SPINAL CONTOURS
FIGURE 2
Note the normal cervical
lordosis, thoracic kyphosis,
lumbar lordosis.

curvature of the spine may be caused by a variety of factors. A difference in leg lengths, pain or muscle spasm due to a tumor, disc degeneration or injury are all causes of nonstructural scoliosis which resolves when the cause of the pain is removed. Diseases affecting the central or peripheral nervous system (poliomyelitis, cerebral palsy) or muscles (muscular dystrophy) may cause scoliosis secondary to spinal muscular imbalance. Myelodysplasia and birth defects of the vertebra may cause a scoliosis. However, the most common form of scoliosis, "idiopathic" scoliosis, has no known etiology and occurs in otherwise perfectly healthy individuals. In addition to the lateral curvature seen with scoliosis, an axial rotational deformity usually occurs (Figure 3). A rotation of the vertebra occurs such that the portion of the vertebra toward the convexity of the curve rotates toward the back and the portion toward the concavity of the curve rotates toward the front. Depending on the severity of the curve there is a variable amount of "hump" (Figures 3 & 6) in which the ribs rotate according to the direction of the vertebra. The cross section of the chest of a patient with severe scoliosis is depicted in Figure 3 indicating that the rotational deformity and distortion of the chest wall caused by scoliosis has led to a significant change in the volume of the chest concavity. The space

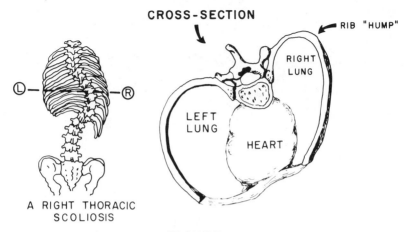

CROSS-SECTION

RIB "HUMP"

RIGHT
LUNG

LEFT
LUNG

HEART

A RIGHT THORACIC
SCOLIOSIS

FIGURE 3

Cross-section through a right thoracic
scoliosis shows distortion of ribs and
diminished volume of right lung.

available for the lungs is diminished. Figure 3 also depicts how a vertebra in severe scoliosis has become distorted and asymmetric.

The term scoliosis refers to sideways or lateral curvature of the spine. Combinations of spinal deformities occur such as lordoscoliosis or kyphoscoliosis. By far the most common spinal deformity and the subject of this review is adolescent idiopathic scoliosis (AIS). This is most commonly a lordoscoliosis with a diminution in the normal amount of thoracic kyphosis.

Physical Findings in Scoliosis

Severe scoliosis is accompanied by an obvious deformity. There is distortion of the chest shape and when present a lumbar curve produces a distortion of flank shape. Figure 5 and 6 depict a severe right thoracic, left lumbar curve in which the shoulders are at uneven height, the right scapula is more prominent, there is a right rib "hump," the chest and head are displaced to the right "out of balance" compared to the lower extremities and there is accentuation of the right flank contour and flattening of the left flank contour. Milder stages of scoliosis have much more subtle physical findings and minimal, if any, deformity.

School screening for scoliosis is intended to detect spinal deformity in its earlier stages. The screening exam consists of observing the height of the shoulders for differences, the height of the scapulae and contour of the hips when standing. The patient is then asked to bend forward (forward bend test) and the examiner sights along the surface of the back looking for thoracic or lumbar "humps" (Figure 6). In the child with AIS this may appear in its earliest form as a slight asymmetry seen only in the forward bend test. Many normal children also have a mild asymmetry of their chest or lumbar musculature without an associated scoliosis. These children will also be screened positively by the forward bend test.

THE NATURAL HISTORY OF SCOLIOSIS

The infantile and juvenile forms of idiopathic scoliosis are extremely rare. Adolescent idiopathic scoliosis usually first appears around the time of puberty and occurs in individuals who are otherwise perfectly healthy. The prevalence of adolescent idiopathic scoliosis is approximately 10-20 per 1,000 but only approximately 1-2 per 1,000 need treatment (Kane & Moe, 1970). Figure 4 summar-

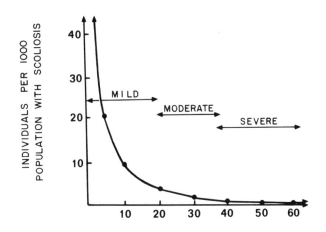

SCOLIOSIS - Severity expressed in <u>degrees</u> of lateral curvature, according to Cobb.

FIGURE 4
Although the incidence of scoliosis is high, scoliosis in need of treatment is infrequent.

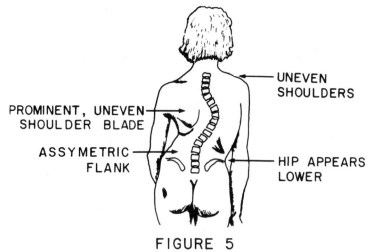

FIGURE 5

Physical findings in severe scoliosis.

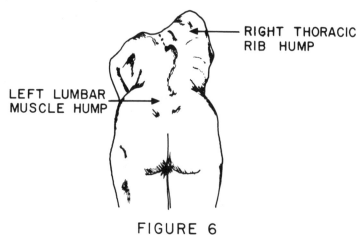

FIGURE 6

"Rib hump" seen on forward bend
test.

izes the prevalence of AIS and indicates that a very high percentage
of the population exhibits a small degree of scoliosis, while only a
very small percentage exhibits severe scoliosis. If scrutinized carefully enough or x-rayed, virtually every normal individual has some

degree of mild scoliosis. The term scoliosis should probably be reserved for lateral curvatures exceeding 10 degrees Cobb. (Figure 7) Mild scoliosis in children occurs with equal frequency in both sexes but progressive, significant AIS occurs at least five times as frequently in females. In the adolescent or child there are no symptoms associated with AIS. Only rarely is AIS associated with pain and a child presenting with both pain and scoliosis requires a careful search for some other etiology for the scoliosis.

The etiology of idiopathic scoliosis remains obscure (Gregori, Pecak, Trontelj, & Dimitrijevi, 1981; Haderspeck & Schultz, 1981; Sahlstrand, 1980; Sahlstrand & Sellden, 1980; Yarom & Robin, 1979). Clearly a genetic predisposition is involved with excellent prevalence studies indicating the hereditary predisposition to AIS (Wynne-Davies, 1968). Offspring of parents with a strong family history have a higher than normal chance of having children with scoliosis but prediction of the frequency or severity of scoliosis among offspring is impossible. Extensive research into the etiology of scoliosis over the past several decades has yielded no primary ab-

A 10° SCOLIOSIS   A 60° SCOLIOSIS

FIGURE 7

Scoliosis measured by the method
of Cobb.

normality. Patients with AIS are taller than and grow later than their normal counterparts. Recent attention has focused on the central nervous system as a possible cause of scoliosis.

Much progress has been made in the last fifteen years in documenting the behavior of untreated scoliosis (Bjerkreim & Hassan, 1982; Clarisse, 1974; Collis & Ponseti, 1969; Fowles, Drummond, L'Ecuyer, Roy, & Kassab, 1978; Nachemson, 1968; Weinstein & Ponseti, 1983). Studies of untreated patients have indicated that many patients with mild scoliosis and many who were past puberty show no worsening of their scoliosis with observation alone. Not rarely spontaneous lessening of mild curvature is seen without any form of treatment. It therefore becomes imperative to distinguish between progressive and nonprogressive AIS. Clarisse's (1974) data and other studies indicate the risk of progression in AIS is much greater in premenarchal girls and in curves over 20 degrees. Typical progressive AIS appears as a small curve (most commonly right thoracic or right thoraco-lumbar) around the time of puberty in the female. Slow progression of the curve occurs until the rapid acceleration of growth just prior to the onset of menses. With acceleration of growth there tends to be rapid worsening of spinal curvature which continues until the end of spinal growth (Duval-Beaupere, 1971). The average for cessation of spinal growth in girls is approximately 16 or three years after the onset of menses but varies among individuals.

At the end of skeletal growth the tendency for rapid progression of AIS stops. The patient with a mild curve (under 20 degrees) can anticipate no further increase in curvature through life. Very severe curves (more than 50 degrees) can be predicted to worsen after the end of growth but at a much slower rate. Data on adult idiopathic scoliosis indicates that this progression after the end of growth may occur at a variable rate. The degree of curvature, imbalance of the torso relative to the pelvis and location of the curve influence the likelihood of progression after the end of growth. Weinstein's (1983) and other data indicate the risk of progression in AIS after the end of growth is very high in curves in excess of 50 degrees, moderate in curves between 30 and 50 degrees and rare in curves less than 30 degrees (Weinstein & Ponseti, 1983). A typical severe curve may increase 1 or 2 degrees per year as an adult, thereby leading to a much more severe curve in later adulthood. Pregnancy before the ages of 25 to 30 may increase the likelihood of progression of existing severe scoliosis (Blount & Mellencamp, 1980). It is postulated that the ligamentous relaxation that occurs with the later stages of pregnancy has a detrimental effect on the spinal ligaments allowing worsening of the curvature.

Studies of untreated adult idiopathic scoliosis indicate a higher incidence of back pain and substantially diminished breathing capacity for adult patients with severe residual scoliosis. In past eras deaths from respiratory compromise due to scoliosis were common. Current studies indicate a higher rate of unemployment, a lower degree of physical activity, lower rate of marriage, and shorter life expectancy among adults with severe untreated AIS.

DOCUMENTATION AND OBSERVATION OF SCOLIOSIS

Radiographs are the standard means for documenting scoliosis. Many pediatricians and orthopedic surgeons have sufficient experience with scoliosis to feel confident in their physical exam that the patient does not have significant scoliosis and choose not to perform an x-ray. A single upright standing x-ray taken from the front or back of the spine is usually sufficient for most AIS. In special situations other views are necessary. A radiograph demonstrates the contour of the vertebra and any distortion in shape as well as the actual degree of lateral curvature of the spine. The standard method of measuring curvature is the angle of Cobb illustrated in Figure 7. Using this method 0 degrees is a perfectly straight spinal segment. With increasing amounts of scoliosis the degree of Cobb angle and curvature increases. In patients at risk for progression of their scoliosis subsequent x-rays are performed with a frequency based on the likelihood of worsening of their curvature. Thus a young child with a severe degree of curvature will be observed very closely with frequent x-rays, while the nearly mature child with a mild curve may not need a subsequent x-ray.

Modern x-ray techniques including high speed film and rare earth screens have drastically reduced the x-ray exposure necessary (DeSmet, Fritz, & Asher, 1981; Gray, Hoffman, & Peterson, 1983). Shielding of the reproductive organs diminishes the dose in this crucial area where possible. Controversy remains as to whether the x-ray is best taken from behind so as to reduce exposure of the breasts or in front so as to reduce x-ray exposure to the bone marrow of the spine. The real risk of harm from diagnostic x-rays is now known, but every attempt to measure this has shown no effect.

Nonetheless, all physicians concerned with scoliosis attempt to minimize x-ray exposure.

This author and others have experimented with non x-ray techniques in the documentation of spinal deformity (Emans, Bailey, & Hall, 1980). Moire photography (Figure 8), a purely photographic technique, documents the contour of the entire back and shows promise for the future as a means of observing scoliosis. Observations indicate that this is sensitive enough to detect changes in progressive AIS. The inclinometer and other physical means of measuring the height of the rib hump on forward bend test have also been used. Most of these techniques however are not widely employed. Since a decision to treat or not to treat scoliosis is largely based upon documentation of progression, it is crucial that x-rays or other documentation be made.

NON-OPERATIVE TREATMENT

Non-operative treatments include exercises, chiropractic treatment, traction, bracing and electrical stimulation of muscles. Non-operative treatment of AIS has as its goal arresting or lessening lateral spinal curvature associated with AIS. Because many milder

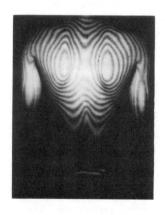 

SLIGHT　LEFT　　　　　　MODERATE　RIGHT
LUMBAR　SCOLIOSIS　　　　THORACIC　SCOLIOSIS
FIGURE 8
Moire' photography for scoliosis.

curvatures spontaneously stabilize or improve without treatment, it is crucial to document progression in milder curvatures before treatment. By so doing many adolescents will be spared unnecessary brace or other treatment. Since the etiology of AIS is unknown, non-operative treatments for AIS are directed at the spinal curvature. Because all non-operative scoliosis treatments are empirical, results must be subjected to rigorous statistical analysis before considering a treatment effective. The last two decades have shown excellent progress in documenting the effectiveness of brace treatment in scoliosis.

Exercises in this author's opinion are a crucial part of any bracing program to prevent stiffness, weakness and increase effectiveness of the brace. Alone exercises are useful in AIS associated with pain, poor posture, in hyperkyphosis or hyperlordosis and in some special circumstances where a great deal of trunk imbalance is involved. Many orthopedists and physical therapists feel exercises alone to be useful in the prevention of progression in AIS but this has not been documented.

The chiropractic profession is enthusiastic about chiropractic treatments for scoliosis. Statistical proof that chiropractic treatment alters the natural history of progressive AIS is lacking. If a child with scoliosis is subjected to chiropractic treatment, it is important to insure objective monitoring of the scoliosis during treatment such as periodic x-rays measured by the method of Cobb. A number of patients have been seen by this author whose AIS progressed during chiropractic treatment beyond the range of bracing treatment and who, if referred earlier in their treatment, might have been spared an operation.

Traction at home or in hospital has achieved popularity in some parts of Europe but has few advocates in this country. Even at those centers where it is used widely in Europe its efficacy is not statistically documented. Brace treatment of scoliosis remains the standard by which all other treatments are measured (Mellencamp, Blount, & Anderson, 1977). There is a long history of attempts to control scoliosis by external support since the time of Hippocrates. The technology of bracing has improved substantially over the last two decades. Results of bracing are more predictable, ranges of curves amenable to bracing are better understood, and braces are more comfortable and better tolerated in day to day activities. Braces can now be fabricated in a shorter time. Since not all AIS is progressive, documentation of progression is important in milder curves before beginning a bracing program. Curves greater than 40 or 45 degrees Cobb have been found to respond poorly in general to bracing and are usually treated surgically instead. Curves under 20 degrees usually do not need treatment. Curves between 20 and 30 degrees which have been proved to be progressive or curves in excess of 30 degrees in individuals in whom growth remains are generally candidates for a bracing program. Exceptions to these guidelines are numerous and depend upon a number of other factors.

Braces push the spine into a straighter position by means of asymmetric pressure on skin, muscle and ribs causing a passive straightening of the spinal curve. Together with exercises an active straightening also occurs. Generally a moderate curve will be reduced by one-third to one-half on a radiograph taken in the brace. With further growth the straightening effect of the brace works in opposition to the tendency of AIS to progress. Braces generally are worn 23 hours a day through the period of skeletal growth with extra time out of the brace for team sports, swimming or exercises. As the end of skeletal growth is achieved, time in the brace is diminished until the brace is finally discontinued.

Many types and names of braces exist. The Milwaukee brace introduced by Blount and Schmidt in the late 1940s has undergone substantial modification over the years but remains the standard to which all other braces are compared. The Boston bracing system and similar braces have been used widely in the last two decades with equal success (Watts, Hall, & Stanish, 1977). Figure 9 depicts the two principal braces used. It is our feeling that a Milwaukee brace or a modification thereof is still necessary for upper thoracic curves, while a brace without a metal superstructure can be used for all curves in the lower part of the thoracic and entire lumbar spine. Individual results of a bracing program depend greatly upon the quality of brace design, fabrication and fitting and a continued vigilance in adjusting the brace. Exercises are necessary to maintain flexibility and enhance the effect of the brace. Since bracing occurs in adolescence, psychological adjustment to the brace is often difficult. A coordinated team approach in which psychological support is available for brace wearers has been used successfully in most scoliosis centers. Scoliosis peer groups composed of patients and parents involved in bracing programs have proved popular. The need for skilled brace makers accustomed to fabricating scoliosis braces, skilled physical therapists and physicians accustomed to scoliosis has led to the development of numerous scoliosis centers for the operative and non-operative treatment of scoliosis.

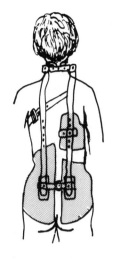 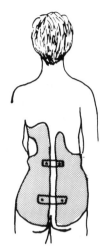

MODIFIED
MILWAUKEE BRACE
FOR HIGH
THORACIC CURVE

BRACE WITHOUT
SUPERSTRUCTURE
FOR
LOW CURVE

FIGURE 9
Types of braces for scoliosis. ·

Figure 10 indicates typical average results of brace treatment of AIS. The average results of bracing are to arrest the progression of AIS. A typical patient shows an approximate 50 per cent improvement in curvature while initially wearing the brace and with further growth and time generally shows mild progression from that initial improvement. At the time of discontinuance of the brace there is generally an increase in curvature. Typically then the patient finishes treatment with approximately the same degree of curvature as his initial curvature. However, approximately 20 per cent of the patients achieve lasting partial correction of their curvature with brace treatment. Some patients with progressive AIS achieve only partial arrest of the progression with bracing and continue to worsen. A small percentage of these patients progress sufficiently to require surgery, either during or at the end of their bracing program. In our recent study nearly 10 per cent of patients started on brace treatment progressed sufficiently to require surgery. Occasionally this is because of poor compliance with the bracing regimen, but there is a

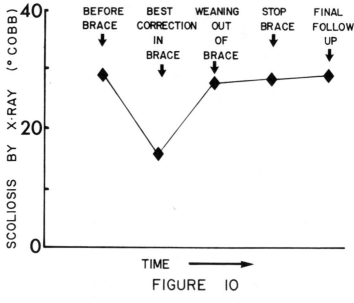

FIGURE   10

TYPICAL   RESULTS   OF   BRACE
TREATMENT   OF   SCOLIOSIS

small number of patients who in spite of complete cooperation continue to do poorly. Since brace treatment is only infrequently cor-
rective, early identification and treatment of progressive AIS is
therefore important.

Bracing has several side effects. Older Milwaukee braces provided distraction of the spine by means of support underneath the
chin. This caused distortion of the jaw and teeth but has been avoided
for the last two decades. A modern Milwaukee brace does not contact the chin or skull. Skin rashes underneath the constant pressure
from the brace are common but generally resolve with modification
of the brace or a change in skin care. Warmer climates provide
more difficulty with skin rashes. Probably the greatest side effect of
brace wear is psychological but this is poorly documented and difficult to measure (Apter et al., 1978). Bracing comes at a particularly
sensitive time in adolescence when body image and peer relations
are crucial. Although many braces can be camouflaged thoroughly
by clothes, some adolescents find the stigma of being different im-
possibly painful. Most adapt completely leading normal existences

with minor, if any, restrictions. Extra time out of the brace is generally granted for swimming, team sports, dance or other physical activities, in an attempt to encourage a normal, active lifestyle. Our experience indicates that a strong, healthy family situation and a well adjusted adolescent will usually comply perfectly with bracing. If there are already other manifestations of behavioral difficulty at the time of initiation of bracing, the chance of successfully complying with a bracing program is greatly diminished. Compliance with bracing seems to hinge on the individual's appreciation that bracing is a necessary and rational treatment to prevent future difficulty. Compliance with bracing seems no different between braces with or without metal neck rings nor according to sex. Professional psychological support and peer groups have been helpful in some instances but many adolescents will refuse to use a brace (Wickers, Bunch , & Barnett, 1977). Often a compromise at part-time bracing is achieved. The progress of a bracing program is monitored with periodic x-rays taken in and out of the brace.

The rationale of the bracing program is to halt the progression of curvature or achieve correction until the end of growth is achieved. This is done in an attempt to avoid either significant deformity at the end of growth or the need for subsequent surgery. Sometimes the chance of success of a bracing program is so small as to not warrant the physical or psychological difficulties involved. Thus very severe curves are often better treated with surgery than with a prolonged course of bracing with little or no chance of avoiding an operation at the end of growth. In this situation surgery is the "conservative" treatment.

Electrical stimulation for the treatment of progressive idiopathic scoliosis has received attention over the last decade. The rationale behind these treatments is that by providing an electrically stimulated contraction of muscles on the convexity of the scoliosis, the spine can be induced to temporarily straighten, halt progression of AIS and possibly achieve correction. Originally this was achieved by means of a surgically implanted stimulator and Bobechko continues this approach which requires an operation (Friedman, Herbert, & Bobechko, 1982). More popular in the United States and Europe is transcutaneous electrical stimulation achieved by means of carbon rubber electrodes applied to the skin with a jelly like conductive medium (Figure 11). Small amounts of pulsating current generated by a portable battery-powered device are passed through the skin into the muscles on the convex side of the curve while the

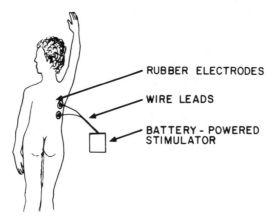

RUBBER ELECTRODES

WIRE LEADS

BATTERY - POWERED
STIMULATOR

Electrodes are worn while sleeping,
muscle contraction stimulated every
6 seconds.

FIGURE II

Transcutaneous electrical stimulation for scoliosis.

patient sleeps at night (Brown, Axelgaard, Nordwall, & Swank, 1980). The author's experience in 60 patients with this device has been encouraging and the available preliminary results of transcutaneous electrical stimulation suggest a success rate in progressive AIS similar to that of bracing. Electrical stimulation for scoliosis remains experimental and longer term studies are needed before it is clear whether this is equivalent to brace treatment. If proven effective, electrical stimulation promises to greatly diminish the psychological burden associated with brace wear during adolescence.

OPERATIVE TREATMENT

Although all the forms of non-operative treatment have as their goal the avoidance of surgery, operative treatment of scoliosis is well tolerated. The advent of safer anesthesia, better blood transfusions and general surgical techniques over the last two decades have made large surgeries safe. The advent of the Harrington Rod apparatus and subsequent spinal surgical developments opened a new era of successful spinal surgical treatment. Before the advent of implantable instrumentation for the spine, patients with scoliosis spent long periods (up to several years) at bedrest in casts waiting for their spinal operation to hopefully heal. Currently most scoliosis operations enable the patient to walk soon after the operation and require at most a light body cast or brace for external support.

Most scoliosis operations have two goals. The first is a fusion or solid stabilization of the spine. Posterior spine fusion is accomplished by pushing aside the muscles overlying the back, removing the surfaces of the small joints of the back, and placing small pieces of bone graft along the spine. This eventually heals in a solid fusion, the equivalent of a single solid bone over the extent of the spine which has been operated upon. In the front of the spine this can be achieved by removing the intervertebral discs and replacing them with bone graft. A fusion thereby solidifies the affected portion of the spine, preventing it from curving further with either subsequent growth or the passage of time. Further progression of scoliosis and its associated difficulties with breathing, pain and deformity are halted.

A second goal of the most current scoliosis operations is partial straightening or "correction" of the existing curvature. This is achieved by one of several metallic surgical implant systems, the most common of which is the Harrington Rod. This extensible bumperjack-like device affixes permanently to the posterior part of the spine and by elongating it stretches the curvature straight. Variations on this are commonly employed and recently more experience has been obtained with segmental spinal instrumentation in which each vertebra is attached to the rod or rods by individual wires. This latter technique has proved very useful in patients with neuromuscular disorders and special situations. Use of Luque spinal segmental instrumentation may entail higher risks of spinal cord injury than traditional Harrington Instrumentation. Its role in surgery for AIS is debated. For some special circumstances instrumentation can be achieved in front of the spine by means of a Dwyer cable and screw type instrumentation and fusion.

A combination of instrumentation and fusion permanently improves and also solidifies the section of the spine operated upon. Depending upon the portion of the spine involved, flexibility is lost. In a patient with a double curve involving both the thoracic and lumbar spine where long segments must be fused, substantial mobility is lost. More long term statistics are needed to know the effects of this loss of mobility. Potential complications of the surgery include failure of a portion of the spine to solidify (pseudarthrosis) requiring repeat operation, post-operative infection, and a very small incidence of paralysis. To diminish the risk of paralysis most centers employ the "wake up test" in which the patient is awakened during the operation enough to cooperate and the function of the spinal nerves checked by asking the patient to move his feet and toes. In some centers electronic spinal cord monitoring has also been used. It is hoped by detecting a paralysis early with these means the spine can be allowed to return to its original position and a permanent paralysis made less likely.

In some very young individuals with severe AIS in whom a great deal of spinal growth is left, there is limited experience with Harrington Rod instrumentation alone without fusion of the spine, delaying fusion to a later time when spinal growth has proceeded further.

Surgical treatment of scoliosis therefore involves some surgical risks and permanent stiffening of the spine. It nonetheless is a highly desirable treatment for more severe curvature, giving the patient a very optimistic outlook for continued good function through life. However, if by early detection, documentation of progression, and non-operative treatment of scoliosis the patient is left with a moderately curved flexible spine, this still is preferable to a straightened operated spine.

ADOLESCENT KYPHOSIS

Adolescent kyphosis, also called Scheuermann's kyphosis or juvenile kyphosis or juvenile epiphysitis, refers to a thoracic hyperkyphosis which occurs more frequently in boys beginning at the onset of puberty. At first this manifests as what appears to be poor posture or round back but unlike postural roundback the patient with Scheuermann's kyphosis is unable to completely straighten the spine on bending backward. With further growth the deformity increases and unlike scoliosis often causes pain during adolescent years. Although milder forms of hyperkyphosis are well tolerated through adulthood, more severe Scheuermann's epiphysitis should be treated aggressively. Some cases may be successfully treated with exercises alone. Others require modified Milwaukee bracing. Although the average result of bracing for scoliosis is arrest of progression of curvature, the outlook is more optimistic in Scheuermann's kyphosis with correction of the deformity much more likely. Although surgical treatment of kyphosis follows many of the same principles of scoliosis surgery, it almost always requires a combined anterior and posterior fusion (Herndon, Emans, Micheli, & Hall, 1981). If noted as part of a school screening program, kyphosis should usually be treated.

CONCLUSION

Much progress has been made in the screening, diagnosis, and treatment of scoliosis over the last several decades. Several unsolved problems remain. Only approximately 1 in 10 school children who are screened as positive for a spinal deformity actually need treatment. The remaining 9 out of 10 individuals with non-progressive AIS usually need no treatment but still are subjected to observation. Often the expense of physician's visits, x-rays and some anxiety is involved in the follow up observation of these non-progressive cases. As yet it is not possible to accurately predict which of these patients detected early will need treatment. It is hoped that ongoing research will determine some predictive factor allowing early distinction between progressive and non-progressive AIS. As yet no real etiology has been determined for AIS and until a causative factor is defined, it is unlikely that accurate predictions about progression will be made. In the meantime every attempt should be made to minimize the impact of observation on those patients with non-progressive scoliosis by diminishing expense and x-ray exposure.

Bracing remains a generally successful treatment for progressive AIS. Some patients fail treatment and it is hoped that a better understanding of the characteristics of these curves will be achieved. The psychological effects on the individual of having to wear a brace for several years during adolescence remain variable. In most individuals it is well tolerated; yet in others it is devastating. More public awareness that scoliosis occurs in otherwise healthy individuals and is not a disease of the severely deformed or crippled will remove some of the social stigma involved with scoliosis brace wearing. It is also hoped that electrical stimulation techniques will prove statistically valid, since some of these techniques could remove much of the physical stigma of brace wearing. Scoliosis centers in which a team approach is used seem to be better accepted by adolescents and peer support groups are sometimes useful in easing the psychological discomfort. School health educational programs discussing scoliosis have helped to make adolescents more aware and accepting of brace wear.

Surgical treatment of scoliosis has improved tremendously over the last several decades and has reached a stage where a scoliosis operation is a predictably safe operation with anticipated good outcome. More research is needed into the long term effects of currently performed scoliosis operations.

Much information about the natural history of untreated scoliosis has been gained. The realization that many milder curves do not worsen or spontaneously improve with further growth has spared many patients unnecessary brace treatment. A better understanding of the natural history of scoliosis in the adult has led to a more reasonable approach to scoliosis surgery. Unfortunately, the current early detection and treatment for scoliosis as well as the ethical dilemma involved in not treating large numbers of patients with progressive AIS probably make is impossible to gain more data in the future on untreated progressive AIS.

REFERENCES

Apter, A., Morein, G., Munitz, H., Tyano, S., Maoz, B., & Wijsenbeek, H. (1978). The psychosocial sequelae of the Milwaukee brace in adolescent girls. *Clinical Orthopedics, 131*, 156-159.

Bjerkreim, I., & Hassan, I. (1982). Progression in untreated idiopathic scoliosis after end of growth. *Acta Orthopedica Scandinavia, 53*, 897-900.

Blount, W.P., & Mellencamp, D. (1980). The effect of pregnancy on idiopathic scoliosis. *Journal of Bone and Joint Surgery, 62A*, 1083-1087.

Brown, J.C., Axelgaard, J., Nordwall, A., & Swank, S.M. (1980). Transcutaneous electrical muscle stimulation for the treatment of idiopathic scoliosis—preliminary results. *Orthopedic Transactions, 4*, 29.

Clarisse, P. (1974). *Prognostic evolutif des scolioses idiopathiques mineures de 10° to 29° en periode de croissance.* Unpublished doctoral dissertation. University Claude Bernard, Lyon.

Collis, D.K., & Ponseti, I.V. (1969). Long term follow up for patients with idiopathic scoliosis not treated surgically. *Journal of Bone and Joint Surgery, 51A*, 425-445.

DeSmet, A.A., Fritz, S.L., & Asher, M.A. (1981). A method for minimizing the radiation exposure from scoliosis radiographs. *Journal of Bone and Joint Surgery, 63A*, 156-161.

Emans, J.B., Bailey, T., & Hall, J.E. (1980). *Preliminary observations in the longitudinal follow up of mild idiopathic scoliosis utilizing shadow Moire topography in Moire fringe topography and spinal deformity.* New York: Pergamon Press.

Friedman, H.G., Herbert, M.A., & Bobechko, W.P. (1982). Electrical stimulation for scoliosis. *American Family Physician, 25*, 155-160.

Fowles, J.V., Drummond, D.S., L'Ecuyer, S., Roy, L., & Kassab, M.T. (1978). Untreated scoliosis in the adult. *Clinical Orthopedics, 134*, 212-217.

Gray, J.E., Hoffman, A.D., & Peterson, H.A. (1983). Reduction of radiation exposure during radiography for scoliosis. *Journal of Bone and Joint Surgery, 65A*, 5-12.

Gregori, C.M., Pecak, F., Trontelj, J.V., & Dimitrijevi, C. (1981). Postural control in scoliosis. A statokinesimetric study in patients with scoliosis due to neuromuscular disorders and in patients with idiopathic scoliosis. *Acta Orthopedica Scandinavia, 52,* 59-63.

Haderspeck, K., & Schultz, A. (1981). Progression of idiopathic scoliosis: an analysis of muscle actions and body weight influences. *Spine, 6,* 447-455.

Herndon, W.A., Emans, J.B., Micheli, L.J., & Hall, J.E. (1981). Combined anterior and posterior fusion for Scheuermann's kyphosis. *Spine, 6,* 125-130.

Kane, W.J., & Moe, J.H. (1970). A scoliosis prevalence survey in Minnesota. *Clinical Orthopedics, 60,* 216-218.

Mellencamp, D.D., Blount, W.P., & Anderson, A.J. (1977). Milwaukee brace treatment of idiopathic scoliosis: Late results. *Clinical Orthopedics, 126,* 47-57.

Nachemson, A. (1968). A long term follow up study of nontreated scoliosis. *Acta Orthopedica Scandinavia, 39,* 466-476.

Sahlstrand, T. (1980). An analysis of lateral predominance in adolescent idiopathic scoliosis with special reference to convexity of the curve. *Spine, 5,* 512-518.

Sahlstrand, T., & Sellden, U. (1980). Nerve conduction velocity in patients with adolescent idiopathic scoliosis. *Scandinavian Journal of Rehabilitation Medicine, 12,* 25-26.

Watts, H.G., Hall, J.E., & Stanish, W. (1977). The Boston brace system for the treatment of low thoracic and lumbar scoliosis by the use of a girdle without superstructure. *Clinical Orthopedics, 126,* 87-92.

Weinstein, S.L., & Ponseti, I.V. (1983). Curve progression in idiopathic scoliosis. *Journal of Bone and Joint Surgery, 65A,* 447-455.

Wickers, F.C., Bunch, W.H., & Barnett, P.M. (1977). Psychological factors in failure to wear the Milwaukee brace for treatment of idiopathic scoliosis. *Clinical Orthopedics, 126,* 62-66.

Wynne-Davies, R. (1968). Familial (idiopathic) scoliosis: A family survey. *Journal of Bone and Joint Surgery, 50,* 24-30.

Yarom, R., & Robin, G.C. (1979). Studies on spinal and peripheral muscles from patients with scoliosis. *Spine, 4,* 12-21.

GLOSSARY

BRACE OR ORTHOSIS. An external device fabricated individually for patient's scoliosis which generally contacts the pelvis and spine.

BOSTON BRACING SYSTEM. A system of brace construction whereby prefabricated brace modules of varying sizes are used to construct an individual scoliosis orthosis. A variety of braces can be constructed using this system including braces with and without metal uprights.

CERVICAL, THORACIC, LUMBAR, SACRAL. Pertaining to the neck, chest, lumbar and sacral regions respectively.

COBB ANGLE. A standardized means of measuring scoliosis on x-rays. 0 degrees is perfectly straight with larger amounts of curvature reflected in larger measured angles.

DISC, INTERVERTEBRAL. A soft tissue structure occupying the space between two adjacent vertebral bodies.

DWYER INSTRUMENTATION. A system of screws and cables implants on the front of the spine to achieve correction of curvature at the same time as fusion is performed.

FUSION. A bony solidification and unification of multiple vertebrae into a single bony unit.

HARRINGTON ROD. A racheted metal rod with hooks on either end designed for implantation adjacent to the posterior spine. Achieves correction through distraction of the curved spinal segment.

HUMP. The prominence noted adjacent to the spine in scoliosis caused by rotation of the muscles or ribs adjacent to the rotated spine.

HYPERLORDOSIS, HYPERKYPHOSIS. An abnormal accentuation of normal spinal contours.

KYPHOSIS. Spinal curvature with convexity backward (posteriorly)

LORDOSIS. Spinal curvature with convexity forward (anteriorly)

LIQUE INSTRUMENTATION. Double rods placed posteriorly on the spine with individual wires passing around the vertebral arches. A form of segmental spinal instrumentation.

MILWAUKEE BRACE, MODIFIED. A modification of the original Milwaukee brace including metal vertical uprights and a ring around the base of the neck. Pressure pads are then attached to the metal uprights.

PROGRESSION. An increase in the degree of curvature.

SCHEUERMANN'S KYPHOSIS (JUVENILE OR ADOLESCENT KYPHOSIS). A progressive kyphosis occurring mostly in males. Vertebra are typically wedged shaped as seen in the lateral x-ray.

SCOLIOSIS. A lateral curvature of the spine, identified as to location, cause, and direction. (A right scoliosis curve is convex to the right.)

STRUCTURAL SCOLIOSIS. Scoliosis which does not disappear or over correct on voluntary bending. The presence of a structural scoliosis implies a permanent deformation of the spinal structures.

VERTEBRA. An individual bone of the spine.

Cigarette Smoking by Adolescent Females: Implications for Health and Behavior

Ellen R. Gritz, PhD

ABSTRACT. Cigarette smoking rates among teenage females have risen progressively since 1968 until surpassing teenage males in 1979. Psychosocial factors underlying the initiation of smoking include peer pressure, adult role modeling and prosmoking messages in advertising, with smoking potentially representing a desired set of personality characteristics. Regular smoking leads to the development of a dependence process, with cessation often difficult to achieve and maintain. Preventing smoking onset is a responsibility of health professionals and educators. Social psychological approaches teaching social skills and techniques for resisting smoking have been effective in reducing rates of initiation in JHS experimental programs.

INTRODUCTION

Cigarette smoking is a behavior with profound biomedical and psychosocial consequences across the lifespan, when examined from perspectives ranging from that of the individual all the way to that of society as a whole. It is a behavior intimately linked with disease and death, which is characterized by those who peddle its material form, the cigarette, in terms of youth, beauty, sexual appeal, personal success and independence. It is a behavior temporally tied to the evolving role of the woman in western society, one which brings greater devastation to our gender just as hard-sought changes are occurring in the realms of education, employment and oppor-

This work was supported by PHS Grant No. 5-R18-CA23974 awarded by the National Cancer Institute, DHHS, and by the Medical Research Service of the Veterans Administration.

An earlier version of this paper was presented at the 90th Annual Convention of the American Psychological Association, August 25, 1982, Washington, D.C.

tunity. And for our gender alone, it carries heavy responsibilities in terms of potential damaging and fatal effects upon the unborn fetus. From the first tantalizing puff by the experimenting adolescent to the last anguished inhalation of the determined-to-quit middle-aged smoker, cigarette smoking may be considered a prototype dependence behavior, an addiction, if you like. For all of these reasons, this topic is of utmost importance to those concerned with the physical health of women, with their social, and with their psychological development across the lifespan.

EFFECTS OF CIGARETTE SMOKING ON HEALTH

The 1980 Report of the Surgeon General, "The Health Consequences of Smoking for Women," summarized the extant knowledge on this topic (USDHHS, 1980). It pointed out in no uncertain language that "women are not immune to the damaging effects of smoking already documented for men,. . .that the first signs of an epidemic of smoking-related diseases among women are now appearing" (USDHHS, 1980, p. v). This delay in morbidity and mortality is primarily due to the later onset of widespread smoking among women than men, dating to World War II rather than World War I. We will see a greater incidence and prevalence of smoking-related disease as women who took up smoking during and since this period, and who continue to smoke, grow older. Without belaboring this point, diseases related to cigarette smoking have been called America's number one preventable health problems.

In the context of this article, it is relevant to summarize briefly the major disease consequences of tobacco use for women (USDHHS, 1980, 1982). Smoking accounts for about 25% of all cancers; it is responsible for 83% of lung cancer cases among men and 43% among women, over 75% in total (American Cancer Society, 1982). It has been causally linked with cancers of the lung, larynx, oral cavity and esophagus. It is considered a contributory factor in the development of bladder, kidney and pancreatic cancer, and has been linked with stomach cancer. The evidence regarding an association between smoking and cancer of the uterine cervix is conflicting and urgently in need of further investigation. In terms of mortality, lung cancer will take the lives of more women in 1983, 34,000, than any other cancer site except breast, for which the estimated mortality is 37,200 (American Cancer Society, 1982). It is predicted that lung cancer mortality will surpass breast cancer mortality nationally some time in this decade (USDHHS, 1982). Indeed, in 1980 female lung cancer mortality surpassed breast cancer mortality in Washington State, a finding which will also occur in California in 1983 (Austin, 1983; Starzyk, 1983). For all sites causally linked to smoking, approximately 53,900 new cases of cancer in women will be diagnosed and 39,750 women will die in this year (American Cancer Society, 1982).

The risk of developing coronary heart disease (including acute myocardial infarction and chronic ischemic heart disease), the major cause of death among both males and females in our country, is increased by at least a factor of two in women who smoke (USDHHS, 1980). Of particular relevance is the ten-fold increase in risk of a myocardial infarction among women who both smoke and use oral contraceptives. This risk is both dose and age-related.

In addition to these two major disease categories, cigarette smoking also contributes to excessive death and disability from chronic obstructive lung diseases (emphysema and chronic bronchitis) and peptic ulcer in women (USDHHS, 1980). The various mechanisms by which smoking interacts with the occupational environment to facilitate development of cancer and other diseases are now directly measurable in women as well as men as we enter an increasingly greater variety of occupations. Potentially differing responses may occur as a function of physiological differences in gender-specific hormonal status. Pregnant women face nine months of exposure to hazardous substances in the workplace, incurring potential teratogenic and perinatal mortality effects (USDHHS, 1980).

Finally, to enumerate the effects of cigarette smoking on reproduction (USDHHS, 1980)—there is a dose-related reduction in birth weight which averages 200 grams, increased risk of miscarriage, spontaneous abortion, various placental abnormalities, preterm delivery, fetal death and neonatal death. Sudden infant death syndrome is more frequent among the infants of smokers (Shannon & Kelly, 1982), as is respiratory illness early in life, and long-term deficits in growth and intellectual development may occur. Finally, impairment in fertility in both men and women has been reported.

BEHAVIORAL ASPECTS OF SMOKING IN WOMEN

Initiation of Smoking

Smoking behavior primarily begins in adolescence, when the disease-connected sequelae seem, at best, vaguely relevant to daily life. Between 1968 and 1979 a series of surveys conducted on 12-18 year olds by the National Clearinghouse for Smoking and Health (NCSH) and the National Institute of Education (NIE) revealed a most disturbing trend in female smoking (NIE, 1979). Not only was it increasing in each age group surveyed at each temporal point, but female smoking now exceeds male smoking, without convincing evidence of downturn.

Overall trends from annual surveys of high school seniors reveal a decline in the percentage of students reporting daily smoking from 28.8% in 1977 to 20.3% in 1981 (Johnston, Bachman, & O'Malley, 1982). This finding is coupled with an increase in the proportion of students who think pack-a-day smoking involves "great risk" to the user (from 51% in 1975 to 64% in 1981). There has been some question about the validity of these downward trends, since data consisted solely of self-report (telephone interview and questionnaire) in all NCSH/NIE and Johnston et al. surveys. When validation measures are taken, such as saliva for thiocyanate analysis, positive self-reports of smoking increase markedly, sometimes by a factor of two (Evans, Hansen, & Mittelmark, 1977; Luepker et al., 1981). The downward trends may be suspect because of a disparity between results from such unvalidated surveys and several validated ones, and because there may now be a bias towards negative self-report with the changing social climate regarding smoking (Johnson, 1982; Luepker et al., 1981).

Nonetheless, when we examine gender-specific figures from the Johnston et al. high school senior survey, the lifetime prevalence of smoking (that is, the overall likelihood of having ever smoked) in 1981 was 73.3% for females and 68.6% for males. For those adolescents reporting any use of cigarettes during the past month, females exceed males—31.6% vs 26.5%—as do they also in the prevalence of those smoking ten or more cigarettes per day, 13.8% vs 12.8%.

Psychosocial Factors in Adolescent Smoking

What has been learned from survey data and research about why teenage female smoking exceeds male, and how may this information be utilized? There are some clues. A survey conducted for the American Cancer Society characterized the teenage female smoker as socially aggressive, sexually precocious, self-confident, rebellious, and rejecting of authority (Yankelovich, Skelly, & White, 1977). Compared to her non-smoking female counterpart, she is more likely to consider parties a favorite leisure time activity, to have a boyfriend, to have had sexual relations, and less likely to feel nervous meeting new people. Compared to the teenage male smoker, she is less socially uneasy, expresses a lesser need to be popular with the opposite sex, and considers smoking less of a social asset.

In general, the literature has characterized adolescent smokers, relative to non-smokers, in the following ways: they tend to hold jobs more frequently while in school; be poorer scholastic achievers (with reduced motivation and lower aspirations); be less likely to go on to college; come more from single parent families; be more likely to have parents or older siblings who smoke; and more likely to have friends who smoke. In addition, teenage smokers are more likely to come from lower income families, to engage in delinquent behaviors, and to be greater experimenters with other drugs such as hard liquor, beer and marijuana (USDHHS, 1980; Wong-McCarthy & Gritz, 1982).

It has been suggested that since adolescents are intensely interested and involved in adapting their actual self-images to their ideal self-images, initiating cigarette smoking may represent the adoption of a set of valued personality characteristics (Wong-McCarthy & Gritz, 1982). In one study, adolescent smokers had real self-concepts and ideal dates more closely matching the stereotypic smoker than did nonsmokers (Chassin et al., 1981). In fact, intentions to smoke among nonsmokers were highest in those adolescents who had self-concepts and ideal dates resembling the smoker stereotype. The ideal image of a woman has radically shifted in this century. In addition to the constant valuation of attractiveness, sexuality and interpersonal facility, there have been added independence, assertion, and career achievement. The ACS survey assessed the reaction of adolescent female smokers to those persons pictured in cigarette advertisements (Yankelovich, Skelly, & White, 1977). Not surprisingly, about half or more saw those persons as attractive, enjoying themselves, well-dressed, sexy, young, and healthy. These are precisely those traits admired by teenage girls, and those which form part of the myth surrounding the social benefits of smoking (Chapman & Fitzgerald, 1982).

The advertising industry is not blind to the female market. A recent article in a trade journal commented that the rise in proportion of women smokers (in Europe), their increased social and economic power, and longer lifespan than men make them ''. . .a prime target as far as any alert European marketing man is concerned. So, despite previous hesitancy, might we now expect to see a more defined attack on the important market segment represented by female smokers?'' (Tobacco Reporter, 1982, p. 6).

While two-page advertisements placed in a major newsmagazine by the Tobacco Institute claim that ''. . .we don't think kids should smoke. Smoking is an adult custom based on mature and informed judgment. That's right, adults, not children'' (Tobacco Institute, 1982). The adult models in cigarette advertisements often are dressed in clothing, and engaged in activities appropriate to adolescents. Such advertisements innundate almost 50% of billboards and a good deal of print advertising. An informal perusal of popular magazines revealed the percentage of all advertising represented by cigarette ads to average around 15%, varying from 7% in Glamour to 29% in Psychology Today (Wong-McCarthy & Gritz, 1982). The prominence and saturation of such advertising may lead teenagers to think of smoking as more popular than it is in actuality. In an attempt to counter pro-smoking messages in media, the American Lung Association distributes a poster, ''Smoking Spoils Your Looks,'' featuring a well-known teenage actress and model making fun of cigarettes (Figure 1). While such antismoking efforts have not been formally evaluated, they do conform to the general principles of social psychological strategies employed in prevention research (see below, Preventing the Onset of Smoking).

A couple of years ago this author hypothesized that ''smoking is just one behavior which may have been 'suppressed' by social norms proscribing appropriate behavior for women in the past, and which now may be 'disinhibited' in a very real sense'' (USDHHS, 1980, p. 286). Watching the data come in, concern for the implications of this statement is greater than ever.

Health Concerns

Culturally induced efforts may influence differences between the way women and men perceive the health consequences of smoking. The 1979 NIE survey showed slightly greater concern among teenage females than males for the health effects of smoking (91.0% vs 85.2%), as well as a greater likelihood to affirm that all cigarettes were equally hazardous (33.7% vs 25.9%). Teenage girls are slightly ahead of boys in the use of ''lower tar'' cigarettes (USDHHS, 1980). Yet, we have the disturbing suggestion that these milder cigarettes are being used as a vehicle for initiation into smoking by females because they produce less of an acute toxic reaction to nicotine exposure (Silverstein, Feld, & Kozlowski, 1980). It is critical that this generation of women remain concerned about smoking-related health issues, yet integrate that concern into motivation and action to quit. Unfortunately, the results of surveys conducted for the Federal Trade Commission (FTC) show substantial ignorance among teenagers for the specific disease consequences of smoking (FTC, 1981). Shortly after the release of the monumental 1979 Surgeon General's Report, one survey showed no gender differences in adult awareness of this Report, but significantly more worry evidenced by female than male smokers over what they knew. Most interesting, "a larger percentage of female smokers attributed a greater health risk from smoking to men than to women, thus suggesting (to these authors) that women's feeling of personal vulnerability may be weaker than men's (Tagliacozzo & Vaughn, 1980, p. 384). It would be interesting to reassess a sample of women on similar issues at this time, when the 1980 Surgeon General's Report has been available for three years and public health efforts have been made to disseminate its vital messages.

PREVENTING THE ONSET OF SMOKING

It is of greater benefit to individuals as well as to society to practice prevention of smoking onset than to engage in remedial cessation efforts: prevention efforts can be more easily aimed at larger audiences than cessation efforts; cessation is difficult to achieve and maintain; any smoking at all may prove harmful to some individuals; and, even if prevention programs only delay initiation for some youth, that delay may have health benefit (USDHHS, 1982).

Adolescent smoking prevention programs have been the focus of a substantial body of research recently (USDHHS, 1982). Seventh grade (junior high school) classes have been the most frequent intervention target groups because of age-related susceptibility and ease of school system access. In general, a pure health education approach is avoided and a social psychological model is utilized. Attention is focused upon immediate negative consequences of smoking (both social and physiological); students are taught about peer pressures to smoke, adult and family role models, and are "innoculated" against prosmoking messages in advertising. Films depicting such social models have been tested with the intent of developing easily transportable interventions (Evans, 1976; Evans, Hansen &

Mittlemark, 1977). Peer leaders have been effectively utilized in teaching prevention curricula (Leupker et al., 1983). Finally, some programs utilize a broad-based approach to life skills training in which developmental issues of adolescence are addressed, e.g., establishing self-esteem and self-confidence, coping with anxiety, and developing autonomy (Botvin & Eng, 1980; Botvin, Eng & Williams, 1980). While there are still major methodological and conceptual issues being debated regarding prevention strategies (Johnson, 1982), experimental programs have achieved approximately a 50% reduction in percentage of students initiating smoking in junior high school, effects that have been shown to last several years (USDHHS, 1982; Luepker et al., 1983).

Some practical recommendations for the physician or other health care professional have been offered (Wong-McCarthy & Gritz, 1982): (1) only cursorily mention the long range health consequences of smoking, since adolescents are generally informed about these; (2) discuss immediate physiological consequences in some detail to highlight acute effects of smoking on the body; (3) discuss alternatives to smoking for promoting desired aspects of self-image among teenagers (e.g., sophistication, maturity, independence); (4) mention the increasing social pressure against smoking, both from legislation and informally; (5) mention negative cosmetic effects of smoking; (6) mention the increasing body of evidence that passive smoking is injurious to nonsmokers; (7) discuss the purported social benefits of smoking, emphasizing that most adults who took up smoking for those reasons now would quit if they were able; (8) help the adolescent develop verbal and behavioral tools for resisting social pressures to smoke; and, (9) have free materials available from the voluntary health agencies (e.g., American Cancer Society, American Lung Association, and American Heart Association) to distribute individually and place in waiting rooms. These strategies will help highlight to the adolescent that smoking is no longer a normative behavior, and should decrease her valuation placed on it for achieving social desirability.

MAINTENANCE AND CESSATION OF SMOKING

Cigarette smoking has recently been boldly identified as a dependence process or addiction. Careful research on pharmacological and behavioral mechanisms in smoking continue to delineate the parameters governing the maintenance and cessation of smoking,

FIGURE 1. THERE'S NOTHING ATTRACTIVE ABOUT SMOKING. . .As illustrated in this poster by Brooke Shields, top teenage model, actress and National Youth Chairman for the American Lung Association. The Shields campaign challenges the idea that smoking is glamorous and proclaims "Smoking Spoils Your Looks." The full color 17x22 poster is available from your local lung association. Smoking Spoils Your Looks. Photo courtesy of the American Lung Association.

and the National Institute on Drug Abuse has published a series of monographs summarizing this work (Gritz, 1980; NIDA, 1977, 1979a, 1979b). The American Psychiatric Association (1980) has included both a Tobacco Dependence Syndrome and a Tobacco Withdrawal Syndrome in the DSM III. Adolescent smoking is so important because 20-30% become regular users by age 18. The spontaneous quit rate in the teen years is estimated at 25% (among regular smokers), with declining probability of cessation as the number of years of regular smoking increases (USDHHS, 1982). Formalized attempts at adolescent smoking cessation programs (as opposed to prevention programs) have fared poorly, with a general lack of interest in participation (USDHHS, 1982).

Considerable difficulty is experienced by many adults attempting cessation, and there is a very high relapse rate, averaging 50-75% in treatment programs (USDHEW, 1979; USDHHS, 1982). About one-third of current regular smokers make a "fairly serious attempt" to quit each year, largely in unaided efforts, according to national survey data. About one-fifth of those who try report success (National Health Interview Survey, 1981; USDHHS, 1980). Adult male smoking has fallen from a prevalence rate of 51.1% in 1965 to 36.7% in 1980, while adult female smoking has only changed from a prevalence rate of 33.3% in 1965 to 28.9% in 1980 (National Health Interview Survey, 1981; USDHHS, 1980). However, recent data provide optimism regarding women's smoking cessation patterns. An analysis of estimated rates of attempted and successful quitting among adult cigarette smokers in 1980 showed that "with respect to the probability of attempting to quit and the success rate, adult men and women cigarette smokers are now indistinguishable" (USDHHS, 1980, p. 26). Between 1978 and 1980, 1.8 million adults quit smoking, 60% of which were female (National Health Interview Survey, 1981). The alleged greater difficulty of women to quit in formalized cessation programs must now be reexamined, and if still found present, researched with regard to differential intervention efficacy (USDHHS, 1980). It is likely that the public health campaign as well as a variety of interpersonal forces are contributing to a breakdown of the former resistance of women to smoking cessation. It is critical to extend this trend, especially in the form of prevention, to our adolescent population.

RESEARCH AND POLICY ISSUES

There are some obvious gaps in our knowledge of teenage female smoking behavior. Primary amongst them is a deeper understanding of what specific constellation of physiological and psychosocial variables lead girls to take up regular smoking. The general set of variables have been outlined, but the weighting factors for each remain to be specified. For example, how important is peer pressure compared to physiological sensitivity to nicotine? How much, if at all, is advertising influential in experimentation with cigarettes or the transition to regular smoking? This kind of information is needed in order to design gender-specific prevention programs which bear consideration because of the difference in male and female prevalence trends. Do teenage girls need to learn about effective weight management, exercise, cosmetic appeal, social skills and assertiveness, instead of assuming that these intangible benefits are delivered along with the tar and nicotine content of a cigarette? What kinds of innovative health education about smoking are effective with adolescent girls: programs and information describing immediate cardiovascular and pulmonary effects, potential reproductive effects, cosmetic effects like skin-wrinkling and tobacco-staining, and the major neoplastic and non-neoplastic disease consequences?

The implication for public policy is clearly to broaden and intensify our national efforts at discouraging initiation in this vulnerable adolescent population. There is little confusion regarding this goal; however, the obstacles to a pro-health policy are well known. The sizeable power of the tobacco industry is manifest in economic, legislative, and political spheres. Fragmentation within the governmental structure between research and programmatic funds also seems to produce an impediment to change (Rittenhouse, 1981). Research on the interactive nature of impediments to policy formation would be valuable. All this will cost money—in research, controlled intervention trials, and finally massive promotional campaigns—but we cannot afford to wait. The ultimate costs in physical and psychosocial morbidity and mortality are incalculable.

REFERENCES

American Cancer Society. (1982). *Cancer facts and figures, 1983.* New York: Author.
American Psychiatric Association. (1980). *Diagnostic and statistical manual of mental disorders* (3rd ed.). Washington, D.C.: Author.
Austin, D. (1983, October 13). Press release for the American Cancer Society. San Francisco.
Botvin, G. J., & Eng, A. (1980). A comprehensive school-based smoking prevention program. *Journal of School Health, 50*(4), 209-213.
Botvin, G. J., Eng, A., & Williams, C. L. (1980). Preventing the onset of cigarette smoking through life skills training. *Preventive Medicine, 9*(1), 135-143.

Chapman, S., & Fitzgerald, B. (1982). Brand preference and advertising recall in adolescent smokers: some implications for health promotion. *Public Health, 72*(5), 491-494.

Chassin, L., Presson, C. C., Sherman, S. J., Corty, E., & Olshavsky, R. W. (1981). Self-images and cigarette smoking in adolescence. *Personality and Social Psychology Bulletin, 7*(4), 670-676.

Evans, R. I. (1976). Smoking in children: Developing a social psychological strategy of deterrence. *Journal of Preventive Medicine, 5*(1), 122-127.

Evans, R. I., Hansen, W. B., & Mittlemark, M. B. (1977). Increasing the validity of self-reports of smoking behavior in children. *Journal of Applied Psychology, 62,* 521-523.

Federal Trade Commission. (1981, May). *Staff report on the cigarette advertising investigation.* Washington, D.C.: Author.

Gritz, E. R. (1980). Smoking behavior and tobacco abuse. In: N. K. Mello (Ed.), *Advances in substance abuse, Volume 1* (pp. 91-158). Greenwich: JAI Press.

Johnson, C. A. (1982). Untested and erroneous assumptions underlying antismoking programs. In: T. Coates, A. Peterson, & C. Perry (Eds.), *Promotion of health in youth* (pp. 149-165). New York: Academic Press.

Johnston, L. D., Bachman, J. G., & O'Malley, P. M. (1982). *Student drug use in America, 1975-1981.* National Institute on Drug Abuse, Department of Health and Human Services, Public Health Service, Alcohol, Drug Abuse and Mental Health Administration, Rockville.

Luepker, R. V., Johnson, C. A., Murray, D. M., & Pechacek, T. F. (1983). Prevention of cigarette smoking: Three-year follow-up of an education program for youth. *Journal of Behavioral Medicine, 6*(1), 53-62.

Luepker, R. V., Pechacek, T. F., Murray, D. M., Johnson, C. A., Hund, F., & Jacobs, D. R. (1981). Saliva thiocyanate: A chemical indicator of cigarette smoking in adolescents. *American Journal of Public Health, 71*(12), 1320-1324.

National Health Interview Survey. (1981, September-October). In: *Smoking and health bulletin.* Department of Health and Human Services, Public Health Service, Office on Smoking and Health, Rockville.

National Institute on Drug Abuse. (1977). *Research on smoking behavior.* NIDA Research Monograph 17, Department of Health, Education, and Welfare, Public Health Service, Alcohol, Drug Abuse and Mental Health Administration, National Institute on Drug Abuse, Rockville: Author.

National Institute on Drug Abuse. (1979a). *Cigarette smoking as a dependence process.* NIDA Research Monograph 23. Department of Health, Education, and Welfare, Public Health Service, Alcohol, Drug Abuse and Mental Health Administration, National Institute on Drug Abuse, Rockville: Author.

National Institute on Drug Abuse. (1979b). *The behavioral aspects of smoking.* NIDA Research Monograph 26. Department of Health, Education, and Welfare, Pubic Health Service, Alcohol, Drug Abuse and Mental Health Administration, National Institute on Drug Abuse, Rockville: Author.

National Institute of Education. (1979). *Teenage smoking, immediate and long-term patterns.* Department of Health, Education, and Welfare, National Institute of Education, Rockville: Author.

Rittenhouse, J. D. (1981). *Children's health and cigarette smoking: Options for public policy.* Paper presented at the 89th Convention of the American Psychological Association, Los Angeles.

Shannon, D. C., & Kelly, D. H. (1982). SIDS and near-SIDS. *New England Journal of Medicine, 306*(16), 959-965.

Silverstein, B., Feld, S., & Kozlowski, L. T. (1980). The availability of low-nicotine cigarettes as a cause of cigarette smoking among teenage females. *Journal of Health and Social Behavior, 21*(4), 383-388.

Tagliacozzo, R., & Vaughn, S. (1980). Women's smoking trends and awareness of health risk. *Preventive Medicine, 9,* 384-387.

Starzyk, P. M. (1983). Lung-cancer deaths: Equality by 2000? (Letter). *New England Journal of Medicine, 308*(21), 1289-1290.

Tobacco Institute. (1982, June 21). Question 3, does cigarette advertising cause kids to start smoking? *Time Magazine,* pp. 66-67.

Tobacco Reporter (Staff). (1982, February). Targetting women. *Tobacco Reporter,* p. 6.

US Department of Health, Education, and Welfare. (1979). *Smoking and health: A report of the Surgeon General.* Department of Health, Education, and Welfare, Public Health Service, Office of the Assistant Secretary for Health, Office on Smoking and Health, (DHEW Publication No. (PHS) 79-50066). Washington, D.C.: Author.

US Department of Health and Human Services. (1980). *The health consequences of smoking for women. A report of the Surgeon General.* Department of Health and Human Services, Public Health Service, Office of the Assistant Secretary for Health, Office on Smoking and Health. Rockville: Author.

US Department of Health and Human Services. (1982). *The health consequences of smoking, cancer. A report of the Surgeon General.* Department of Health and Human Services, Public Health Service, Office on Smoking and Health. Rockville: Author.

Wong-McCarthy, W. J., & Gritz, E. R. (1982, August). Preventing regular teenage cigarette smoking. *Pediatric Annals, 11*(8), 683-689.

Yankelovich, Skelly, & White, Inc. (1977). *A study of cigarette smoking among teen-age girls and young women.* Summary of the Findings. Department of Health, Education, and Welfare, Public Health Service, National Institutes of Health, National Cancer Institute, (DHEW Publication No. (NIH) 77-1203).

The Troubled Teen:
Suicide, Drug Use, and Running Away

Barbara Sommer, PhD

ABSTRACT. The incidence and predisposing factors of suicide, drug use, and running away in adolescence are described with particular attention to teenage females. A description of common underlying themes and of steps for prevention and treatment is included.

The common thread tying together suicide, drug use, and running away is self-destruction. All three have received considerable attention in the past decade. The second characteristic linking them is their illegal or quasi-illegal status. Finally, all three have been associated with disturbed family relationships and feelings of isolation and alienation. The purpose of this paper is to provide current factual information about these potentially self-destructive behaviors with regard to their incidence and predisposing factors among adolescents, with particular attention to young women.

SUICIDE

Two levels of suicide will be addressed, completed suicide and attempted suicide.

Completed Suicide

There is no single reason why young women commit suicide. The immediate or precipitating causes vary greatly. Shneidman (1976, pg. 5) states that suicide involves "an individual's tortured and tormented logic in a state of inner-felt, intolerable emotion." In her work with self-poisoning, McIntire (1980) refers to problems associated with the 5 Ps: parents, peers, privation, punctured romance, and pregnancy. Suicide is seen by many researchers as an act trig-

The author is grateful to Ilana Davis for her comments on the manuscript.

gered by intense feelings of anger and rage resulting from some loss—of a loved one or of an important status, or perhaps even of a sense of previously-experienced well-being (Toolan, 1981). It is a rare occurrence prior to puberty, although the previous assumption that children do not commit suicide is seriously questioned by some researchers (Cohen-Sandler, Berman, & King, 1982; Toolan, 1981).

While current statistics provide a general picture of the incidence of completed suicide, the figures grossly underestimate the true number. Actual incidence rates are estimated to be 2-3 times that reported (Seiden, 1969; Toolan, 1962, 1975). Much of the underreporting stems from the stigma associated with suicide. Many suicide statistics are probably hidden in accident figures (Anderson, 1981; Cohen-Sandler et al., 1982; Toolan, 1981). Accidents, particularly motor vehicle accidents, are the leading cause of death in adolescents.

Suicide rates are *higher* for whites than nonwhites, are *lower* for persons under 19 years of age than for older age groups, and are *higher* for males than for females, as shown in Figure 1 (Vital Statistics, 1978).

The rate of suicide at all ages has increased over the past fifteen years (Figure 2). Worldwide, suicide rates are higher in industrialized and more prosperous societies (Teicher, 1979). Rates are lower in time of war and higher in periods of economic depression, although this trend is less marked among adolescents (Holinger & Offer, 1981).

Following the release of the Surgeon General's report (Health, United States 1980, 1983), there was a flurry of media publicity concerning the dramatic rise in suicide and accident rates among the young. Figures 1 and 2 illustrate that rise, but also show that the rate of increase differs considerably by gender. It is the increase in male suicide that is particularly dramatic.

Methods used. Looking at the methods used, the majority of both sexes used firearms. The difference is in the second ranking with females being more likely to ingest poison, e.g., a drug overdose, and males more likely to be in the hanging category.

Attempted Suicide

There is little agreement as to whether those who commit suicide and those who attempt it without actually killing themselves represent two discrete groups or whether they range on a more quantita-

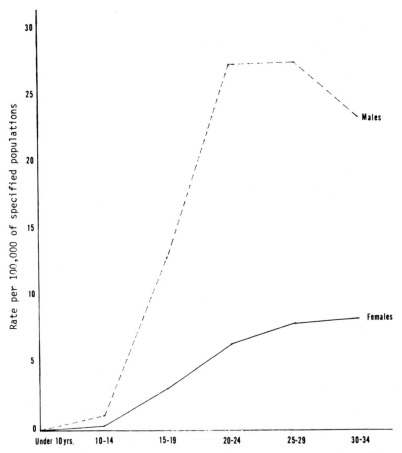

Figure 1.   Death rates by age group, for females and males, 1978
(Vital Statistics of the United States, 1978).

tive continuum reflecting intent and lethality. Attempts to distinguish among persons who have threatened, attempted, or completed
suicide, have not been useful in either risk assessment or treatment
and prevention. An alternative typology is Shneidman's (1966)
fourfold classification system of lethality:

High. The person definitely wants to die and anticipates that
his or her actions will result in death.
Medium. The person is ambivalent, playing a partial, covert or
unconscious role, e.g., through carelessness or drug abuse.

Low. The person plays a small, but significant role, e.g., takes risks.
Absent. No lethal intent.

Garfinkel, Froese and Hood (1982) offer a typology of three categories in addition to completed suicide:

> *Suicide attempts.* Deliberate acts of nonfatal self-injury where conscious intention is to die.
> *Suicidal gestures.* Conscious intention is not to die but to draw attention to experienced conflict and distress.
> *Suicidal ideation.* Where self-destructive thoughts are seldom translated into behavior.

Because of a lack of consensus on the presence or absence of distinctions between suicide attempts and suicidal gestures, both are subsumed in the category of attempted suicide for the purposes of this paper. Angle (1980) asserts that ". . .most self-destructive behavior by adolescents is of low lethality" (Angle, 1980, pg. 24).

Incidence

Estimates of attempted suicide range as high as 45.6 per 100,000 adolescents (Teicher, 1979). The ratio of attempted to completed suicide has been estimated in a range from 16:1 to 200:1 (McIntire, 1980; Peck, 1982). The precise demographics are unknown.

Based on available figures, it appears that more females than males attempt suicide. Petzel and Cline (1978) estimated that females attempted suicide 2-3 times as often as males, and Garfinkel et al., (1982) reported a ratio of 3 female adolescents to 1 male seen for attempted suicide in a metropolitan hospital emergency room over a period of seven years. However, for those under 13-1/2 years of age, the sex ratio was nearly equal. Attempts recorded for 215 teenagers at the Pittsburgh Poison Center showed that among 13-15 year olds, the ratio was 7 females to every male, and among 16-18 year olds, it was 3 to 1 (Moriarty, 1980).

A contributing factor to the greatest proportion of females than males being treated for attempted suicide may be connected with sex roles. Adolescent females on the average are more willing to seek help and are more self-revealing than males (Sommer, 1976). These differences may influence who is likely to turn up at a clinic for as-

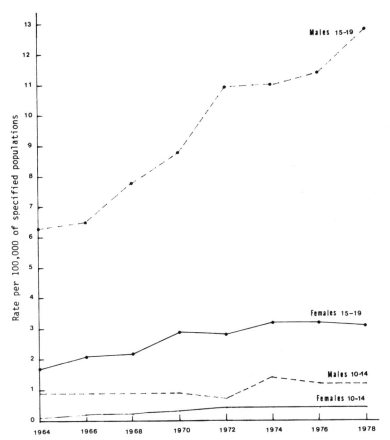

Figure 2. Death rates for suicide, for females and males, ages 10-14 years and ages 15-19 years, at 2-year intervals from 1964 to 1978 (Vital Statistics of the United States, 1964-1978).

sistance or contact suicide prevention personnel. Gender differences may also influence the degree to which an attempt is considered an accident and may also influence the judgment of hospital or clinic personnel in accepting such accounts, e.g., risky behavior by a teen-age male being attributed more often to bravado.

Methods used. Poisoning is the most common type of suicide at-tempt (Cohen-Sandler et al., 1982; Garfinkel et al., 1982; McIntire, 1980; Tishler, McHenry & Morgan, 1981). Of the 505 attempted suicides studied by Garfinkel et al., (1982) 88 percent of the group

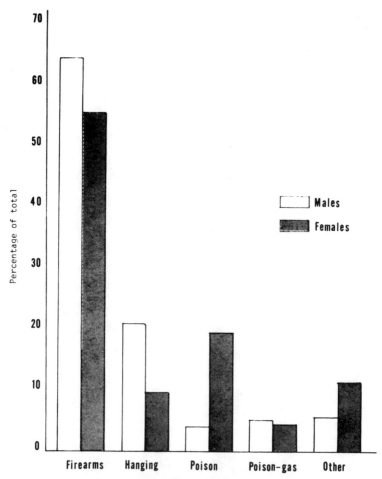

Figure 3.   Methods of completed suicide, males and
females, ages 10-19 years, 1978. (Vital
Statistics of the United States, 1978).

had overdosed on drugs, usually household pain relievers. Only
small percentages used other means—wrist laceration (8%), hanging (2%), and jumping from heights or in front of automobiles
(2%). A breakdown by method and sex was not given. In a sample
of 108 emergency room treatments of adolescents, Tishler et al.
(1981), found that 87 percent of the females and 74 percent of the
males ingested drugs or other toxic substances. Cohen-Sandler et al.
(1980), found that in their hospitalized sample, only males had attempted suicide by either hanging or jumping.

Recurrence

A major concern in cases of attempted suicides is the likelihood of a subsequent attempt. There does not appear to be a simple means of judging, nor is there much information about recurrence rates. Angle (1980) reported on a 6-24 month follow-up of 40 female and 10 male adolescents who had attempted suicide by self-poisoning. She located 26 of the original 50 and their background factors appeared representative of the entire original group (the breakdown by gender in the final group was not given). Of these 26, using an *a priori* assessment based on a structured interview when first seen and the circumstances of the original event, six were judged at low risk for subsequent lethal attempts, and twenty at high risk. At the time of follow-up, none of the low risk adolescents had made further suicide attempts, while eight (40%) of the high risk group made repeated attempts. She also noted that in three other studies of self-poisoning, 26-52 percent of the adolescents had made one or more prior suicide attempts.

Garfinkel et al. (1982) found, in a follow-up seven years later, that eight of the suicidal group had died: 4 from drug overdose, 3 in motor vehicle accidents, and 1 from hanging. The five deceased persons from the control had all died from disease. In contrast, Rauenhorst (1972) did not find more evidence of subsequent disturbance when contrasting a group of 50 young women, ages 16-30 years who had attempted suicide with a matched sample approximately 1-1/2 years later. The generalization from these few findings is that for adolescents, a suicide attempt means being at risk for a subsequent one, but not necessarily so.

Predisposing Factors

A number of underlying factors have been associated with suicide (Holinger & Offer, 1981). While suicide shows a familial association, the evidence for genetic factors is weak, with greater than chance occurrence in specific families attributed to interpersonal dynamics and conditioning or learning.

Psychiatric factors associated with completed and attempted suicide have involved themes of depression, loss, or being an expendable child, i.e., unwanted and rejected by family members. Mental illness, particularly depression and psychosis (schizophrenia), are frequently related to suicide (Peck, 1982).

Earlier views of mental disturbance held that depression did not occur in the young. Getting around this almost ideological premise, researchers argued for "masked depression" or "depressive equivalents" (Teicher, 1979; Toolan, 1981). These terms cover behaviors which range from temper tantrums to boredom. Such a broad conceptualization lacks face validity in that frenzied activity, restlessness or boredom are not part of the general diagnosis of depression, nor are they unique to suicidal adolescents. Current evidence suggests that classic depressive symptoms do occur even among prepubertal children—symptoms of unpleasant mood, weight loss, sleep difficulty, poor appetite, and thoughts of suicide (Pfeffer, 1981).

Teicher (1979) has indicated that suicide in adolescence is the end result of an extensive history of social and familial instability characterized by recurring losses of love objects and the existence of few meaningful relationships. Incidence of divorce and separation is higher in the families of suicidal adolescents (Adams, Lohrenz, Harper & Streiner, 1982; Cohens-Sandler et al., 1982; Garfinkel et al., 1982; Tishler, McHenry & Morgan, 1981; Teicher, 1979; Toolan, 1981; Wenz, 1979).

Garfinkel et al. (1982), comparing emergency room reports for 505 suicidal children and adolescents with a matched nonsuicidal sample found that more of the suicidal youngsters had a history of alcohol and drug use. There were some religious differences with Catholics and Jews slightly under-represented and Protestants slightly over-represented in the suicidal group, relative to controls. The suicidal group had families characterized by higher rates of mental illness, alcohol or drug abuse, and medical illness. More of the suicidal youth had histories of psychiatric and medical illness and contacts with psychosocial services. Cohen-Sandler et al. (1982), also found more alcohol and drug use among the parents of suicidal youngsters than among parents of a nonsuicidal comparison group. Comparing life stress scores of suicidal youngsters with those of comparison groups of depressed (but not suicidal) and psychotic youngsters, they found that the suicidal group showed a history of increasing amounts of stress accompanying maturation. Overall, the suicidal group experienced more chaotic events and associated losses in the area of family relationships than the other two groups. There also continues to be evidence of a higher than chance incidence of depression and suicide attempts by others in the families of suicidal youngsters (Garfinkel et al., 1982; Pfeffer,

1981; Shaffer, 1974; Teicher, 1979; Tishler, McHenry & Morgan, 1981).

There is disagreement on the role of socioeconomic status with some researchers claiming that rates are the same across social class (Farberow & Schneidman, 1961) and others disagreeing, reporting a higher suicide rate among persons of lower socioeconomic states (Finch & Poznanski, 1971). A 1971 review of the literature revealed a greater frequency of suicide in urban areas; higher rates in shifting, rather than stable, communities; higher rates in overcrowded conditions, and among youth in foster homes or boarding schools; and for youngsters with a history of secondary school absenteeism and juvenile delinquency (Finch & Poznanski, 1971). Anderson (1981) found that many suicidal adolescents were not enrolled in school at the time of the attempt.

Married teenagers have a higher suicide rate than unmarried ones, a reversal of the pattern in older age groups. Divorced teens have an even higher rate of suicide than married ones (Dublin, 1963; Petzel & Cline, 1978; Seiden, 1969). First-born children may be over-represented among adolescent suicides (Cohen-Sandler et al., 1982; Lester, 1966).

There are occasional incidents of contagion—where suicide attempts increase in imitation of a media figure or by adolescents imitating peers. Kenney and Krajewski (1980) reported a flurry of suicide attempts in a school where one youngster had received considerable attention and hospital visits from teachers and friends. Such incidents of contagion like other forms of hysteria, are probably self-limiting. Finch and Poznanski (1971) stated that contagion is more common when small groups of emotionally-disturbed adolescents are in close contact with one another.

Treatment and Prevention

A broad spectrum of treatment modalities has been developed covering family therapy (Fishman & Rosman, 1981; Richman, 1981), hospitalization (Kenney & Krajewski, 1980), behavioral therapy (Keefe & Ward, 1981), and crisis centers (Peck, 1982). Suggestions for prevention range from strengthening family bonds and community institutions to individual psychotherapy and developing the reality testing abilities of adolescents.

Prediction and assessment of risk continue to be problematic (Beck, Resnick & Lettieri, 1974; Holinger & Offer, 1981). The classic danger signs associated with suicide in adolescence are (Schneider 1977):

1. Loss of interest and loss of drive in social environment;
2. Feelings of sadness and emptiness;
3. Eating disturbances, usually anorexia nervosa;
4. Sleeping disturbances, mostly insomnia;
5. Feelings of loneliness.

DRUGS

As with the distinction between suicide gestures and suicide attempts, the distinction between drug use and drug abuse is controversial. One could argue that the first-mentioned behaviors in these two pairs do not fall within the definition of self-destruction. Because of the lack of agreement on a clear demarcation, suicidal gestures were subsumed within the category of attempts. In a similar sense drug abuse is subsumed within the larger category of drug use in this review. Because of that, we cannot infer a specific incidence of self-destructive behavior, but rather, gain a rough gauge of number at risk. This gauge is based on an assumption of continuity along a dimension from use to abuse, again recognizing that the one does not necessarily lead to the other.

The most comprehensive and current overview of drug use by adolescents in the U.S. is that provided by Johnston, Bachman and O'Malley (1981) as part of their Monitoring the Future project in which large samples of graduating high school seniors are surveyed each year. Data are now available from 1975 to 1981. These studies, supported by the National Institute on Drug Abuse, provide a general picture of normative adolescent drug involvement, against which specific individuals and groups can be compared. The nature of the sample does exclude school dropouts, a group characterized by a higher rate of drug use (Johnston, 1973).

Incidence

In the class of 1981 (based on a sample of 18,267 seniors from 109 public and 9 private high schools across the U.S.), 66 percent acknowledged at least some illicit use of a drug at some time, and that estimate is probably a conservative one (Johnston, Bachman & O'Malley, 1981).

Specific Drugs

Figure 4 illustrates the prevalence and recency of use of eleven specified drugs. *Alcohol* is the most common drug used, followed by *cigarettes* and *marijuana* with *stimulants* in fourth place. The *stimulant* category, which was designed to reflect use of drugs such as amphetamines, is probably inflated due to the senior's reporting of over-the-counter preparations for diet control and staying awake.

Sex Differences

Sex differences remain, though diminishing. A greater proportion of males is involved in drugs, and daily use is heavier (see Table 1). Males' daily use of *marijuana* is twice that of female adolescents.

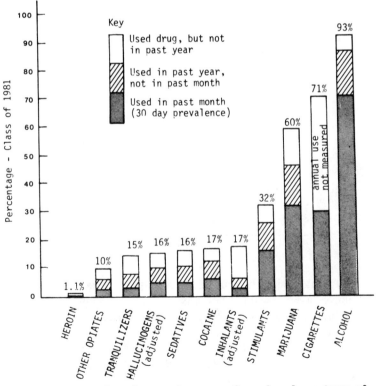

Figure 4. Prevalence and recency of use for eleven types of drugs, Class of 1981 (Johnston, Bachman & O'Malley, 1981, pg. 19).

Table 1. Percentage of females and males using
          specified drugs within the past 30 days,
          Class of 1981 (Johnston, Bachman & O'Malley,
          1981, page 26).

Females	Sex Males	Drug type
27.3	35.3	Marijuana
1.1	1.9	Inhalants*
0.6	2.2	Amyl/Butyl Nitrites
2.6	4.6	Hallucinogens*
1.4	3.4	LSD
1.0	1.7	PCP
5.0	6.3	Cocaine
0.1	0.3	Heroin
1.8	2.4	Other Opiates
16.7	14.7	Stimulants
3.9	5.2	Sedatives
2.4	2.9	Barbiturates
2.4	3.7	Methaqualone
2.6	2.7	Tranquilizers
65.7	75.7	Alcohol
31.6	26.5	Cigarettes

* Unadjusted for known underreporting of certain drugs.

Only two categories show more female use. One is *stimulants*, but as mentioned earlier, that finding is probably confounded by reports of diet pill use. The other difference is that more females smoke *cigarettes* (see Gritz, this issue).

Despite these sex differences, about the same proportions of both sexes report using some *illicit drug other than marijuana* during the past year (32% of males and 34% of females, with the proportions at 25% males and 22% females, with *stimulants* removed). Earlier surveys also show similar patterns of drug use for female and male adolescents (Beschner & Treasure, 1979). Frequent use of alcohol is disproportionately concentrated in males and they drink more in a single setting. Corroborating information is provided by a second series of studies done in 1974 and 1978 on a nationwide sample of high school students for the National Institute on Alcohol Abuse and Alcoholism (First Statistical Compendium, 1981). Table 2 shows the breakdown by gender. Frequency of alcohol use was stable for those two years.

A similar survey of a representative sample of households across the U.S. found that among 12 to 17 year olds, 39% of the 1,128 males and 36% of the 1,037 females drank in the past month. Marijuana use for the same period was reported by 19% of the males and 14% of the females (Fishburne, Abelson & Cisin, 1980).

Socio-demographic Variables

The subject sample for the class of 1981 comes from four regions of the U.S. For the group, drug use is highest in the Northeast, followed by the West, North Central, and South, in decreasing order, with the exception of many more students in the West using cocaine and eschewing PCP, Methaqualone, and cigarettes.

There is a general correlation between population density and drug use with the incidence being higher in large metropolitan areas, particularly for cocaine and marijuana. Rates of drug use also differ within specific population groups. College-bound adolescents show a lower rate with the exception of alcohol use which is about the same. Other studies have shown ethnic and racial differences with higher rates of drug use for blacks, Hispanics, and Native Americans (with some tribal variability), especially for drugs other than marijuana. There is a lower than average rate for Asian-Americans. However, the connection between ethnicity and drugs is not a direct and simple one, and there is a trend toward increasing homogeneity across population groups (Gersick, Grady, Sexton & Lyons, 1981). One remaining difference is that black and Hispanic youth tend to progress from alcohol to marijuana to heroin and cocaine, while white youth insert an intermediate step from marijuana to psychedelics and barbiturates, in the progression to cocaine and heroin (Gersick et al., 1981).

Time trends. Looking at the patterns since 1975, there is no change in age of first experience for *illicit drugs other than mari-*

Table 2. Adolescent drinking: Percentages of 10th-12th graders in each drinking level by sex (Rachal et al., 1981).

Drinking level	1974 Study		1978 Study	
	Males (N=2820)	Females (N=3115)	Males (N=2276)	Females (N=2642)
Abstainers	15.5	23.3	22.7	27.3
Infrequent	8.1	14.2	6.2	9.1
Lighter	16.0	20.9	15.7	22.0
Moderate	17.0	17.5	15.8	17.4
Moderate/Heavier	21.7	15.7	18.8	15.7
Heavier	21.7	8.4	20.9	8.9

juana. Rather, the mix of drugs has changed over the years, i.e., *hallucinogens* and *sedatives* are down and *stimulant* use is up. Overall, with the exception of the increasing use of cocaine, illicit drug use is remaining fairly stable.

Predisposing Factors

Factors which have been shown to precede the initial use of a drug are (1) low academic performance, (2) crime, (3) low self-esteem, (4) depressive mood, and (5) rebelliousness and other behavior characteristics such as behavior disorders and aggression in childhood (Kandel, 1981; Clayton, 1981). These are qualities which we intuitively associate with drug abuse, as well as use. A common distinction is *use* as representing infrequent and limited intake, with *abuse* as "the illegal use of a controlled substance, or use of drug in a manner or to a degree that leads to adverse personal or social consequences" (Strategy Council on Drug Abuse, 1973). It is also possible that different factors affect drug use and drug abuse. For example, whether one has *ever tried* a drug is likely to be influenced by availability, whereas *habitual use* will be affected not only by availability, but by motivational factors and reinforcement systems.

There are many theories of the etiology of drug abuse ranging from sociological explanations to psychodynamic interpretations and covering virtually all perspectives between the two. Lettieri, Sayers and Pearson (1980) have compiled an anthology of perspectives—over thirty theoretical points of view grouped into four general categories: (1) *theories on one's relationship to oneself,* e.g., personality deficiency, cognitive control, existential issues, ego/self-assessment, coping; (2) *theories on relationship to others,* e.g., social deviance, self-esteem, developmental stages, drug subcultures; (3) *relationship to society,* e.g., social influences, learned behavior, social control; and (4) *relationship to nature,* i.e., biological factors—genetic theory, metabolic deficiency, opiate receptors, etc. The generalization suggested by the existence of all these theories is that no single explanation holds for a majority of drug abusers.

Some earlier generalizations have been called into question. In general, socio-demographic factors are of little predictive value, and the existing few differences are disappearing. Findings are contradictory regarding a connection between psychopathology and drug abuse. There is little consensus regarding the role of personality factors such as low self-esteem, locus of control, sensation-seeking, etc. The only personality factor showing a consistent relationship with drug use is unconventionality-deviance or rebelliousness (Gersick et al., 1981; Kandel, 1980).

In the past more psychodynamic, family-oriented and interpersonal explanations were offered for female drug abusers with more subcultural deviance-oriented variables suggested for males, but most current researchers claim gender differences are too small and subtle to justify distinct explanations and programs (Gersick et al., 1981).

The age-old controversy of family versus peer influence continues. In his review Glynn (1981) argues that neither one nor the other is solely influential at any point, and rejection of family influence at one level does not imply rejection at another. Currently, adolescents tend to be more influenced by *peers* regarding marijuana, but less so for other illicit drugs. Rather than a general drug culture, the data support a view of specific network of peers oriented toward specific drugs (Jessor & Jessor, 1977). There is a strong relationship between an adolescent's drug use and that of friends, especially for alcohol and marijuana. In addition, parental use of hard liquor predicts adolescent drinking and use of drugs other than marijuana (Kandel, 1980).

Prevention and Treatment

Hanson (1980) after a comprehensive review of drug education programs, concluded that they have not been an effective deterrent to drug use. While these programs are successful in increasing knowledge, they have not resulted in decreased drug use.

Huberty (1980) predicts that treatment programs will continue along current lines. The treatment of choice for adolescents continues to be family therapy. Specialized hospital-based treatment programs are on the increase. An example is the St. Mary's Adolescent Chemical Dependency Unit in Minneapolis. Techniques of "confrontation and support, insight, and information, meditation and relaxation, and identification of a unique value system will comprise the programming" (Huberty, 1980, p. 312).

Parent groups have developed and currently there is an upsurge in youth groups who in turn are forming networks with adult groups. Examples are S.A.D.D. (Students Against Driving Drunk) and Y.F.D.A. (Youth for Drug Free Alternatives). The National Institute on Drug Abuse has many pamphlets and publications for the development of such groups. In addition Resnick and Gibbs (1981) describe a variety of peer program approaches appropriate both within and outside the school setting.

RUNNING AWAY

As with suicidal behavior and drug use, running away from home may be a highly self-destructive act, or a slightly risky one, or simply a brief outing with little serious intent or consequence. Concern about the potential hazards of running away and the high incidence of adolescents leaving home led to three self-report studies. A major one was the *National Statistical Survey on Runaway Youth* (Opinion Research Corporation, 1976) which covered 14,000 households across the United States. A *runaway* was defined as "a youth between the ages of 10 and 17 inclusive, who has been absent from home, at least overnight, without parental or guardian permission" (ORC, 1976, p. 3).

Brennen, Huizinga and Elliott (1978) made a detailed analysis of that study as well as two others—the OYD study (Office of Youth Development, 1975) which involved a sample of 2,400 households producing 44 runaway youth, and the HEW-Colorado study (Brennen, Blanchard, Huizinga & Elliott, 1975) of an institutional sample of 139 runaways. All three studies surveyed nonrunaways for comparison.

Incidence

Based on the *National Statistical Survey,* an estimated 733,000 youths ran away in 1975, 53% were male and 47% female, with the largest proportion being 16 years of age (31%). Runaways utilizing federally-supported runaway shelters have a modal age of 15 and are more likely to be female (Nye and Edelbrock, 1980). Most runaways do not go far, nor do they stay away for a long time. The survey showed that 40 percent were away only one day and another 30 percent returned within the week. Adolescents who leave home are more likely to come from low-income families, and the lowest rate of running away is in the middle-income strata. Regionally, the highest rate is in the Northwest and West-Midwest (5.0% and 4.1% annually) and lowest in the Northeast and Southeast (1.5% and

2.1%). From this survey, it was estimated that one youngster in eight will run before his or her eighteenth birthday (Nye & Edelbrock, 1980). However, this estimate appears high in light of Brennen et al.'s (1978) analysis of the three studies showing that the percentage of runaway youth ranged from 1.7 to 1.9 percent of the national youth population ages 10 through 17 years.

Based on data from interviews with a representative sample of adolescents, ages 12-17, the National Health Center Statistics in 1975 revealed that 8.7% of the females and 10.1% of the males claimed to have run away from home on at least one occasion. The higher estimate was probably due in part to the broader definition of running away, one that did not include staying away overnight—"leaving or staying away on purpose, knowing you would be missed, intending to stay away from home, at least for some time" (Justice & Duncan, 1976).

Predisposing Factors

In the combined findings of the OYD and HEW-Colorado studies, half of the runaways indicated that arguments with parents or problems at home were the reasons for running away, and 60 percent of these youngsters claimed the problems were of a long-term nature. There were no differences in these percentages across gender, age, ethnic group, or social class.

The second most frequently cited reason was a personal problem such as loneliness, depression, or pregnancy. Again, there were no general differences by gender and age groupings, although personal problems were cited by 49 percent of the upper socio-economic group and by 35 percent of the lower socio-economic group. A majority (66%) claimed that the problems were long-standing. Other reasons given were problems with friends, (14%), school (14%), and police (9%). Similarly, in the *National Statistical Survey* the runaway youths viewed themselves as having problems in relationships with parents, teachers, and peers, in addition to having low academic aspirations and poor self-image.

The three self-report studies revealed a number of different types of runaway—for example, delinquent lower-class runaways, alienated middle-class delinquent runaways, middle-class normal youth runaways, etc. The point of multiple motives and types of runaway is evident in the literature with some studies showing that running away is an indicator of severe emotional problems and connected with other illicit behaviors such as truancy, alcohol abuse, and delinquency. Repeated running away is listed among the criteria for nonaggressive conduct disorders in the *Diagnostic and Statistical Manual, III* (DSM-III, 1980, pp. 48-49). In contrast, other studies focus on running away as a healthy response to a difficult home situation or as an indicator of adventurousness. There is also the category of "throwaway"—youth categorized as runaway but who in fact have been overtly rejected by their families (Adams & Gullotta, 1983).

A number of typologies have been developed, and one consensus is that there is a segment of runaway youth who appear neither disturbed, nor disadvantaged, nor socially incapacitated. These youngsters more often than not have made an unplanned exit, often in response to an argument, and generally return home within a week, and do not run away again. Another group, fortunately less numerous, are runaways who are, additionally, involved in delinquency, have poor school performance, and poor peer relations. These are the repeaters who travel far and stay away a long time. In their case the running away is less a central issue and more an indicator of serious problems at home, at school, and with social relationships (Edelbrock, 1980).

Edelbrock (1980) found a high incidence of running away among a sample of 2,967 disturbed children and adolescents, ages 4-16 years, who had been referred for mental health services. The incidence of running away was much higher than that in a comparison sample of 1,300 non-referred youth. Among the females, ages 12-16 years (N = 472), characteristics associated with running away were truancy, use of alcohol and drugs, screaming, suicidal talk, and attempted suicide. These were also significantly associated with running away in males of the same age. Sexual problems, poor school work, and disobedience at home also correlated with running away for these disturbed young women.

Hinted at, but less often assessed, is the possibility of incest, which is likely to be of more salience among adolescent females. While males may run in reaction to incestuous wishes, females are more likely to be actual victims of incest. Nilson (1981) makes the point that when an older girl in a family runs away, those seeking to assist her should be alert to the possibility of incestuous abuse. Reilly (1978) studied fifty female teens brought before the Boston Juvenile Court for running away. The young women showed more truancy, hostility, and incidents of lying or cheating, relative to a comparison group of nonrunaway, nondelinquent teens. In psychiatric interviews, 14 of the 50 indicated that sexual tension with their fathers or stepfathers precipitated their flight, and one had experienced direct sexual advances. While on the run, many of the sample had experimented with drugs and had engaged in sexual activity.

Nilson underscores the similarity of motives which are often common to running away, suicide, and drug abuse—a deep longing for peace, longing for someone one hopes will be nurturing and loving, and a strong underlying rage toward an unloving, abusive, or absent parent (Nilson, 1981).

Treatment and Prevention

As with other self-destructive behaviors, the wide range of etiology requires a multiplicity of treatment alternatives—family counseling, individual therapy, crisis centers, and shelters. Brennen et al., (1978) listed a number of early warning signs of running away:

A. Family context
   1. Withdrawal of love by parents
   2. Parental remoteness and disinterest
   3. Disorganized or ineffective discipline practices
   4. High conflict over freedom and autonomy
B. School context
   1. Loss of aspiration and involvement
   2. Academic failure
   3. Perceived denial or access to desirable school roles
C. Peer context
   1. High levels of runaway behavior among friends
   2. High levels of antisocial attitudes and behavior among friends
   3. High levels of normative peer pressure toward anti-social behavior
   4. Large amounts of time spent with friends coupled with minimal time spent with parents
   5. Loneliness and boredom (especially among older [16+] females)

Treatment suggestions are currently predicated on some sort of typology, since the data show considerable heterogeneity among runaway youth. Examples are Orten and Soll's (1980) treatment typology based on an assessment of the degree of alienation between home and child. Brennen et al., (1978) have offered a more complex typology with specific treatment recommendations by type, for example, for "Class II, rebellious and constrained middle-class dropout girls" they suggest individual counseling directed at anger and hostile feelings, family therapy, school counseling and contact with a more positive peer culture. The *Journal of Family Issues,* Vol. 1, No. 2, June 1980, is devoted in its entirety to the problem of running away and contains sources for prevention and treatment programs.

SUMMARY AND CONCLUSIONS

With respect to *suicide,* while the rate of completed suicide for young women is less than that of young men, both sexes are clearly at risk, especially in the later teen years. Further, if the statistics are correct, more female than male teens attempt suicide. Associated factors are feelings of anger and depression and a history of disturbed family relationships, particularly parental loss.

Studies of *drug use* show parental behavior being imitated by offspring, particularly with regard to drinking. Peer influence seems particularly strong with respect to marijuana use. Female adolescents use drugs less heavily and less frequently than males, but the general pattern of use is similar. A wide variety of etiological factors have been hypothesized with current emphasis on situational ones. Drug education has not met its promise in prevention. More emphasis is being placed on the development of support groups among both parents and teens aimed at reducing use, along with continued use of clinic treatment for drug dependency.

Running away occurs with nearly equal frequency among females and males, although incest is more likely a problem for females. Disturbed family relationships coupled with poor school performance, and time spent with anti-social peers who themselves run, markedly increases the likelihood of an adolescent's running away from home. As with the other two self-destructive behaviors, family therapy as well as work with the individual teen is strongly warranted.

There are a number of ways in which we can begin to deal with the self-destructive adolescent. One is in the area of primary prevention, of developing resources for improving parenting—clinics, education, and improved opportunities for observation. At present there are only about 20 to 30 infant programs in the U.S. extensively involved with emotional and social development as well as assessment of health and cognitive factors (Turkington, 1983). Given the small family size common today, many parents-to-be have had no experience with babies. Increased exposure to infants in the high school and late adolescent years as well as exposure to models of good parenting would be of educational value. The opportunities inherent in the media for education in this regard remain to be exploited.

Other steps are to strengthen community support systems. Schools, social organizations, religious groups, etc. can help reduce the costs of inadequate family life by providing opportunities for attachment and the building of skills. Such community groups may provide a buffer against social disorganization on a larger scale. With proper staffing they can be a haven of acceptance, consistency, and predictability.

Finally, treatment programs aimed at developing the inherent strengths of adolescents may offset early deficits. A theme running through virtually all theories of adolescent development is one of a strong internal push for independence and competence. In developing those qualities the lack of self-esteem, feelings of helplessness, and lack of control so characteristic of self-destructive teens may be at least remedied in part.

REFERENCES

Adams, G. R., & Gullotta, T. (1983). *Adolescent life experiences.* Monterey, CA: Brooks/Cole.

Adams, K. S., Lohrenz, J. G., Harper, D., & Streiner, D. (1982). Early parental loss and suicidal ideation in university students. *Canadian Journal of Psychiatry, 27,* 275-281.

Anderson, D. R. (1981). Diagnosis and prediction of suicidal risk among adolescents. In C. F. Wells & I. R. Stuart (Eds.), *Self-destructive behavior in children and adolescents* (pp. 45-59). New York: Van Nostrand Reinhold.

Angle, C. R. (1980). Recurrent adolescent suicidal behavior. In M. S. McIntire & C. R. Angle (Eds.) *Suicide attempts in children and youth* (pp. 24-30). Hagerstown, MD: Harper & Row.

Beck, A. T., Resnick, H. L. P., & Lettieri, D. J. (1974). *The prediction of suicide.* Bowie, MD: Charter Press.

Beschner, G. M., & Treasure, K. G. (1979). Female adolescent drug use. In G. M. Beschner & A. S. Friedman (Eds.), *Youth drug abuse* (pp. 169-212). Lexington, MA: Lexington Books.

Brennan, T., Blanchard, F., Huizinga, D., & Elliott, D. (1975). *Final report: The incidence and nature of runaway behavior.* Boulder, CO: Behavioral Research and Evaluation Corp.

Brennan, T., Huizinga, D., & Elliott, D. S. (1978). *The social psychology of runaways*. Lexington, MA: Lexington.

Clayton, R. R. (1981). The delinquency and drug use relationship among adolescents: A critical review. In D. J. Lettieri & J. P. Ludford (Eds.), *Drug abuse ad the American adolescent* (pp. 82-103). NIDA Research Monograph 38, U.S. Department of Health and Human Services. Washington, D.C.: U.S. Government Printing Office.

Cohen-Sandler, R., Berman, A. P., & King, R. (1982). Life stress and symptomatology: Determinents of suicidal behavior in children. *Journal of the American Academy of Child Psychiatry, 21,* 178-186.

DSM-III; *Diagnostic and statistical manual of mental disorders*, third edition (1980). Washington, D.C.: American Psychiatric Association.

Dublin, L. J. (1963). *Suicide: A sociological and statistical study*. New York: Ronald Press.

Edelbrock, C. (1980). Running away from home: Incidence and correlates among children and youth referred for mental health services. *Journal of Family Issues, 1,* 210-228.

Farberow, N. L., & Shneidman, E. S. (Eds.) (1961). *The cry for help*. New York: McGraw-Hill.

Finch, S. M., & Poznanski, E. O. (1971). *Adolescent suicide*. Springfield, IL: Charles C. Thomas.

First statistical compendium on alcohol and health. (1981). National Institute on Alcohol Abuse and Alcoholism, Department of Health and Human Services. Washington, D.C.: U.S. Government Printing Office.

Fishburne, P. M., Abelson, H. I., & Cisin, I. H. (1980). *National survey on drug abuse: Main findings 1979*. National Institute on Drug Abuse. U.S. Department of Health and Human Services. Washington, D.C.: U.S. Government Printing Office.

Fishman, H. C., & Rosman, B. L. (1981). A therapeutic approach to self-destructive behavior in adolescence: The family as patient. In C. F. Wells & I. R. Stuart (Eds.), *Self-destructive behavior in children and adolescents* (pp. 292-307). New York: Van Nostrand Reinhold.

Garfinkel, B. D., Froese, A., & Hood, J. (1982). Suicide attempts in children and adolescents. *American Journal of Psychiatry, 139,* 1257-1261.

Gersick, K. E., Grady, K., Sexton, E., & Lyons, M. (1981). Personality and sociodemographic factors in adolescent drug use. In D. J. Lettieri & J. P. Ludford (Eds.), *Drug abuse and the American adolescent* (pp. 39-56). NIDA Research Monograph 38. U.S. Department of Health and Human Services. Washington, D.C.: U.S. Government Printing Office.

Glynn, R. J. (1981). From family to peer: Transitions of influence among drug-using youth. In D. J. Lettieri & J. P. Ludford (Eds.), *Drug abuse and the American adolescent* (pp. 57-81). NIDA Research Monograph 38. U.S. Department of Health and Human Services. Washington, D.C.: U.S. Government Printing Office.

Hanson, D. J. (1980). Drug education; Does it work? In F. R. Scarpitti & S. K. Datesman (Eds.), *Drugs and the youth culture* (pp. 251-282). Beverly Hills, CA: Sage.

Health, United States 1980 (1983). DHHS Publication No. (PHS) 81-1232. U.S. Department of Health and Human Services. National Center for Health Statistics. Washington, D.C.: U.S. Government Printing Office.

Holinger, P. C., & Offer, D. (1981). Perspectives on suicide in adolescence. *Research in Community and Mental Health, 2,* 139-157.

Huberty, D. J. (1980). Treating the young drug user. In F. R. Scarpitti & S. K. Datesman (Eds.), *Drugs and the youth culture* (pp. 283-315). Beverly Hills, CA: Sage.

Jessor, R., & Jessor, S. L. (1977). *Problem behavior and psychosocial development—A longitudinal study of youth*. New York: Academic Press.

Johnston, L. D. (1973). *Drugs and American youth*. Ann Arbor, MI: Institute for Social Research.

Johnston, L. D., Bachman, J. G., & O'Malley, P. M. (1981). *Student drug use in America 1975-1981*. National Institute on Drug Abuse. U.S. Department of Health and Human Services. Washington, D.C.: U.S. Government Printing Office.

Justice, B., & Duncan, D. (1976). Running away: An epidemic problem of adolescence. *Adolescence, 11,* 365-371.

Kandel, D. B. (1980). Drug and drinking behavior among youth. *Annual Review of Sociology, 6,* 235-285.

Kandel, D. B. (1981). Drug use by youth: An overview. In D. J. Lettieri & J. P. Ludford (Eds.), *Drug abuse and the American adolescent* (pp. 1-23). NIDA Research Monograph 38. U.S. Department of Health and Human Services. Washington, D.C.: U.S. Government Printing Office.

Keefe, F. J., & Ward, E. M. (1981). Behavioral approaches to the management of self-destructive children. In C. F. Wells & I. R. Stuart (Eds.), *Self-destructive behavior in children and adolescents* (pp. 309-327). New York: Van Nostrand Reinhold.

Kenney, E. M., & Krajewski, K. J. (1980). Hospital treatment of the adolescent suicidal patient. In M. S. McIntire & C. R. Angle (Eds.), *Suicide attempts in children and youth* (pp. 70-86). Hagerstown, MD: Harper & Row.

Lester, D. (1966). Sibling position and suicidal behavior. *Journal of Individual Psychology, 22,* 131-132.

Lettieri, D. J., Sayers, M., & Pearson, H. W. (1980). *Theories on drug abuse: Selected contemporary perspectives.* NIDA Research Monograph 30. U.S. Department of Health and Human Services. Washington, D.C.: Government Printing Office.

McIntire, M. S. (1980). The epidemiology and taxonomy of suicide. In M. S. McIntire & C. R. Angle (Eds.), *Suicide attempts in children and youth* (pp. 1-23). Hagerstown, MD: Harper & Row.

Moriarty, R. W. (1980). Suicide attempts and gestures: The poison center perspective. In M. S. McIntire & C. R. Angle (Eds.), *Suicide attempts in children and youth* (pp. 87-91). Hagerstown, MD: Harper & Row.

National Center for Health Statistics (1975). *Self-reported health behavior and attitudes of youths 12-17 years, United States. Vital and Health Statistics,* PHS Pub. No. 1000-Series 11-No. 147. Washington, D.C.: Government Printing Office.

Nilson, P. (1981). Psychological profiles of runaway children and adolescents. In C. F. Wells & I. R. Stuart (Eds.), *Self-destructive behavior in children and adolescents* (pp. 2-43). New York: Van Nostrand Reinhold.

Nye, I. F., & Edelbrock, C. (1980). Some social characteristics of runaways. *Journal of Family Issues, 1,* 147-150.

Office of Youth Development (OYD) (1975). *Runaway youth: From what to where?* Conference of OYD, Kansas City MO.

Opinion Research Corporation (1976). *National statistical survey on runaway youth* (Part I). Princeton, NJ: ORC.

Orten, J. D., & Soll, S. K. (1980). Runaway children and their families. *Journal of Family Issues, 1,* 249-261.

Peck, M. (1982). Youth suicide. *Death Education, 6,* 29-47.

Petzel, S. V., & Cline, D. W. (1978). Adolescent suicide: Epidemiological and biological aspects. *Adolescent Psychiatry, 6,* 239-266.

Pfeffer, C. R. (1981). The distinctive features of children who threaten and attempt suicide. In C. F. Wells & I. R. Stuart (Eds.), *Self-destructive behavior in children and adolescents* (pp. 106-120). New York: Van Nostrand Reinhold.

Rachal, J. V., Maisto, S. A., Guess, L. L., & Hubbard, R. A. (1981). Alcohol use among youth. In National Institute on Alcohol Abuse and Alcoholism. *Alcohol consumption and related problems.* Alcohol and Health Monograph No. 1. Rockville, MD: the Institute, in press.

Rauenhorst, J. M. (1972). Follow-up of young women who attempt suicide. *Diseases of the Nervous System, 33,* 792-797.

Reilly, A. (1978). What makes adolescent flee from their homes? *Clinical Pediatrics, 17,* 886-893.

Resnik, H. S., & Gibbs, J. (1981). Types of peer program approaches. In National Institute on Drug Abuse, *Adolescent peer pressure: Theory, correlates, and program implications for drug abuse prevention* (pp. 47-89). U.S. Department of Health and Human Services. Washington, D.C.: Government Printing Office.

Richman, J. (1981). Family treatment of suicidal children and adolescents. In C. F. Wells & I. R. Stuart (Eds.), *Self-destructive behavior in children and adolescents* (pp. 274-290). New York: Van Nostrand Reinhold.

Schneider, L. (1977). Adolescents and suicide. In B. L. Danto & A. H. Kutscher (Eds.), *Suicide and bereavement* (pp. 163-175). New York: Arno Press.

Seiden, R. H. (1969). Suicide among youth: Review of the literature 1900-1967. *Bulletin of Suicidology,* Supp.

Shaffer, D. (1974). Suicide in childhood and early adolescence. *Journal of Child Psychology and Psychiatry, 15,* 275-291.

Shneidman, E. S. (1966). Orientations toward death. In R. W. White (Ed.), *The study of our lives.* New York: Atherton Press.

Shneidman, E. S. (Ed.) (1976). *Suicidology: Contemporary developments.* New York: Grune & Stratton.

Sommer, B. (1976). *Puberty and adolescence.* New York: Oxford University Press.

Strategy Council on Drug Abuse (1973). *Federal strategy for drug abuse and drug traffic prevention.* Washington, D.C.: U.S. Government Printing Office.

Teicher, J. D. (1979). Suicide and suicide attempts. In J. D. Noshpitz (Ed.), *Basic handbook of child psychiatry,* Vol. 2. New York: Basic Books.

Tishler, C. L., McHenry, P. C., & Morgan, K. C. (1981). Adolescent suicide attempts: Some significant factors. *Suicide & Life Threatening Behavior, 11,* 86-92.

Toolan, J. M. (1962). Suicide and suicidal attempts in children and adolescents. *American Journal of Psychiatry, 118,* 719-724.

Toolan, J. M. (1975). Suicide in children and adolescents. *American Journal of Psychotherapy, 29,* 339-344.

Toolan, J. (1981). Depression and suicide in children: An overview. *American Journal of Psychotherapy, 35,* 311-322.

Turkington, C. At risk. *APA Monitor,* June 1983, pg. 6.

Vital Statistics of the United States, 1978, Vol. II-Mortality, Part A (1978). National Center for Health Statistics, Vital Statistics of the United States. U.S. Department of Health and Human Services. Washington, D.C.: Government Printing Office.

Wenz, F. (1979). Sociological correlates of alienation among adolescent suicide attempts. *Adolescence, 14,* 19-30.